AF506004

The Rise of the Japanese Specialist Manufacturer

The Rise of the Japanese Specialist Manufacturer

Leading Medium-Sized Enterprises

Ferguson Evans

palgrave
macmillan

First published 2008 by
PALGRAVE MACMILLAN

Palgrave Macmillan in the UK is an imprint of Macmillan Publishers Limited, registered in England, company number 785998, of Houndmills, Basingstoke, Hampshire RG21 6XS.

Palgrave Macmillan in the US is a division of St Martin's Press LLC, 175 Fifth Avenue, New York, NY 10010.

Palgrave Macmillan is the global academic imprint of the above companies and has companies and representatives throughout the world.

Palgrave® and Macmillan® are registered trademarks in the United States, the United Kingdom, Europe and other countries.

ISBN 13: 978–0–230–21842–0   hardback
ISBN 10: 0–230–21842–3   hardback

This book is printed on paper suitable for recycling and made from fully managed and sustained forest sources. Logging, pulping and manufacturing processes are expected to conform to the environmental regulations of the country of origin.

A catalogue record for this book is available from the British Library.

A catalogue record for this book is available from the Library of Congress.

10   9   8   7   6   5   4   3   2   1
17   16   15   14   13   12   11   10   09   08

Printed and bound in Great Britain by
CPI Antony Rowe, Chippenham and Eastbourne

Contents

Aline

List of Figures

List of Tables

List of Boxes

List of Abbreviations

AABI	Association of Asian Business Incubation
ANIEs	Asian New Industrial Economies
AGM	Annual General Meeting
ASEAN	Association of South East Asian Nations
ASTEM	Advanced Software Technology and Mechatronics Research Institute of Kyoto
FDI	Foreign direct investment
GATT	General Agreement on Tariffs and Trade
IMF	International Monetary Fund
ISO	International Standards Organization
JANBO	Japan Association of New Business Incubation Organizations
JASDAQ	Japan Association of Securities Dealers Automated Quotations
JCH	Japan Company Handbook
JETRO	Japan External Trade Organization
JODC	Japan Overseas Development Corporation
JTSB	Japan Trustee Services Bank
KSK	Kaigai Shinshutsu Kigyo Soran (Survey of Companies Investing Overseas)
LME	Leading Medium-sized Enterprise (Japanese: *chuken kigyo*)
MBO	Management buy-out
MCI	Ministry of Commerce and Industry
METI	Ministry of Economy Trade and Industry
MITI	Ministry of International Trade and Industry
MKB	Mijojo Kigyo Ban (Unlisted Company Edition)
MTB	The Master Trust Bank of Japan
NIE	Newly industrializing economy
Nikkei	Nihon Keizai Shinbun (Japan Economic Journal)
OECD	Organization of Economic Cooperation and Development
OECF	Overseas Economic Cooperation Fund
PCB	Printed circuit board
SMEs	Small and Medium-sized Enterprises
TQC	Total quality control
WTO	World Trade Organization

Acknowledgements

My interest in the Japanese economy goes back some 30 years and I have met and benefited from the insights of many individuals far too numerous to mention – Chinese, Korean, British, American and French as well as, needless to say, Japanese. When I started seriously studying the subject and directing my attention more specifically to specialist manufacturers and what I have come to refer to as leading medium-sized enterprises I was fortunate to gain from the help and advise of Roger Leigh, then at Middlesex University. It was during that time also that I was graciously received by quite a few businessmen of Japanese firms, notably in Japan and Taiwan, who were kind enough to give of their valuable time to discuss the way their firms operated. These included Sakamoto Chikashi and Mumo Nobuyuki of Pentel, Awai Shiro of Mabuchi, Wakino Yoshikazu and Murakami Kazuhiko of Shofu, Ohdaira Hiroshi and Inami Tatsuo of Union Tool, and Sodeyama Yukio and Ikeda Osamu of Tanaka Kikinzoku. I am grateful to Union Tool for permitting me to reproduce a table in Chapter 9. In addition, I also grate-fully acknowledge Shimano Yoshizo, the current chairman of Shimano, for granting me permission to reproduce in Chapter 11 quotations from his recent book.

This book has been written in Singapore. It would have been a duller preoccupation without the crowd at the Wagon Wheel. Unfair though it may seem to single out individuals, I feel bound to mention Tony who *always* poured a good pint, Boh who *occasionally* drove me home under inclement conditions, and Koh who *invariably* drove me up the wall whatever the weather.

Finally, I must thank my wife Aline, to whom this book is dedicated, for putting up with all this.

1
Introduction

The time is the beginning of the 1980s. Six of us are seated along a rectangular table, three on each side, in a Japanese restaurant on the other side of the world from Japan. At one end of the table is a Japanese trade official talking in English to the second-in-command of the host country's agency tasked with coaxing inward direct investment sitting opposite him. Diagonally across at the other end of the table is the executive of a Japanese manufacturer talking across to me, but in Japanese. Facing each other between these two pairs are the director of the agency and the president of the said manufacturer. Linguistic incompatibility has rendered them silent. In no uncertain terms, and making no attempt whatsoever to mute his tones, the executive is launching on a tirade addressed to me but aimed at the trade official. He detested that type. They just sat on their behinds and did nothing. They had no idea what work really meant. The message was not new to me. I had had first-hand experience with a number of small businesses in Japan where I had lived by that time for over a decade. I was familiar with the Mercedes-driving machine shop owner quivering with indignation as he recounted the frosty reception given him by an administrative flunky in ill-fitting trousers or a bank clerk half his age and commanding a tiny fraction of his income.

What did surprise me, however, was the blatant aggressiveness of the verbal assault. Given the silent couple in the middle, there was virtually no chance that the official could not have overheard. It would seem the executive had elected to use me as a sound box for deflecting his antagonism to its intended destination without engaging in – by just a whisker – direct confrontation. This magnified one of the central themes of this book, and in fact this small group epitomized the actors and the stage in the narrative which is about to unfold. First there was the trade official, representative of central authority and command in a nation state, who is here designated as the 'samurai'. Then there was the executive, or the 'artisan', with his localized predilections sometimes but by no means always at variance with attitudes emanating from the center – and usually not so openly hostile as in the incident just cited. The president embodied the specialist manufacturer, further defined here as the leading medium-sized enterprise because, although playing a key role in the economy, it remains distinct from the elite business 'samurai' fraternity. The host-country agency personnel stood for the internationalization and ultimately globalization which the president was looking to see his specialist firm embark upon. Finally, there was me. I was the interpreter, a role I am again assuming here.

The background

Over a quarter of a century has passed since that incident. For Japan this period of 25 years can be divided into four phases: boom, bubble, recession, and recovery. The boom was the result of a successful economy producing superior products and breaking into new markets. It was so successful in fact that it fueled prophecies of the coming of the Japanese century, Japan assuming the baton from a wilting United States. This still held for a time as the yen doubled in value against the dollar from 1985 to 1987. Exports suffered maybe, but the economy was in full stride, its firms vigorously making good the disadvantage with a foreign direct investment binge and thereby further endorsing the integrity of its challenge. But integrity of a different kind was about to be put to the test and to be found wanting. By the end of the decade the economy was brimming over with cash of no fixed abode. So it went into Van Goghs, gold-flaked ice-cream and – most obdurately – real estate. Property prices soared, generating mouth-watering collateral to purchase more property based on mounting debt supposedly secured by real assets whose visibility was rapidly receding over the horizon. When the membrane of credibility finally nicked solid fact the bubble burst. Banks were plunged into an ocean of bad debts, the depth and magnitude of which they were only to admit to reluctantly and fitfully as the subsequent decade wore on. The recession was to grind on into the first years of the new millennium and, for some, Japan even now remains off the radar screen of economic relevance.

This does not make sense for a number of reasons. To begin with, Japan still has the world's second largest economy. As of the end of 2005, at $4,534 billion Japan had a gross domestic product over twice that of China ($2,234 billion) and India ($806 billion) combined (The Economist, 2007a). As more and more workers of these far more populous nations trade in shovels for machines, of course, this situation will change dramatically over the next ten years. Productivity and per capita income will take considerably longer though. At any event, by 2003 Japan's economy was turning the corner and on the road to recovery. A random sampling of indicators for that year and the four following years help to substantiate this. In the final quarter of that first year, Japan's economy had its strongest growth for 13 years, expanding at an annualized rate of seven percent, urged on by once-more robust exports. By the mid-term of fiscal 2004, the operating profits of companies listed on the stock exchange had risen by over 30 percent year-on-year, while by 2005 the country's banking regulator felt able to announce that the 16-year bad-loan crisis was officially over. Added to which, full-time employment was growing faster than part-time employment for the first time in over a decade. In January 2006, loans by the overseas branches of Japanese banks were the most for 16 years, spurred on by the flurry of investment activity by Japanese firms in Asia, Eastern Europe and Russia. Back home it was being foreseen that steel production would be at its highest likewise for 16 years, goaded by revitalized demand from automobiles and ship-building, and on the way to achieving the highest output of crude steel for 30 years by 2008. By 2007, meanwhile, equipment investment by leading companies in most of the industrial sectors had been on the increase for five consecutive years.

Another reason is globalization and regional integration within that. Coincidentally perhaps the term 'globalization' was coined just around the time when the artisan was maligning the samurai over a sushi repast while the president of the leading medium-sized enterprise was mulling manufacturing beyond his native shores. Globalization as it was being fashioned by accelerated communications and transportation was presenting this president with fresh options for expanding his business. These options were not confined within national institutional and structural bounds. They offered to the capable of any nation the chance to act on a larger stage. Or, in other words, globalization invited those factors of any national economy with particular competitive advantages to step forward. Manufacturing was Japan's outstanding forte, especially by then in machinery and electronics. Moreover, although the Japanese economy was to supposedly stumble and meander thereafter, manufacturing overall continued to upgrade its productivity, right through the recession and into the recovery.

Japanese manufacturing also continued to expand abroad through exports and direct investment in overseas plants. Manufacturing in proximity to markets was a strategic option already manifested in Japanese affiliate plants in North America as early as the 1960s and spreading to Western Europe over the following two decades or so. However, it could also be brought on by globalization pressures. Threatening Western protectionist posturing coupled with cheaper and increasingly capable manpower induced many a Japanese manufacturer to set up abroad. But here proximity took on a different significance. For the bulk of this foreign direct investment found its way to the neighboring countries of the East Asian region – the Asian newly industrializing economies (ANIEs) of Hong Kong, Singapore, Taiwan and South Korea in the 1970s, the ASEAN 4 of Thailand, Malaysia, Indonesia and the Philippines in the 1980s, China in the 1990s and after. Proximity was the key here because it facilitated integration of manufacturing on a regional basis. To a considerable extent also, it allowed for Japanese manufacturers to insinuate themselves, frequently as instigators, into embryonic industries as they took form within the countries in the region. Many became the indispensable feeders of progress. This is one very good reason why, contributions by local Japanese manufacturing affiliates aside, as of 2005 Japan was the leading source of imports for China, Taiwan, Thailand, Indonesia, and South Korea, and the second most important source for Malaysia and the Philippines (*The Economist*, 2007a).

This leads to the final reason, which is the Japanese specialist manufacturer within globalization and regional integration. One of Japan's most impressive achievements since the 1980s has been the way it has managed to mask a fair measure of its supremacy. Not a little of the reason for this can be found in the specialist manufacturer. Of course, there are end-user specialist makers like Nintendo who are eminently visible to a wide global audience, and they are part of the story here. But many more are not. They are the manufacturers of intermediate components and materials, of machine tools and other devices indispensable for the manufacturing of parts, subunits, and final products, and of the machines that make the machines. They produce miniaturized electronic motors, ultra-precise cutting machines, and super-hard drills, failing which downstream production would be seriously compromised if not rendered completely impossible. That is to say, many of them have cornered niche monopolies or at least quasi-monopolies which constitute what are called chokepoints. Alternatively they can be seen as core strengths which can have implications like the following.

***

At around 10:35 on the morning of 16th July, 2007 an earthquake of magnitude 6.8 shattered the calm of the Chuetsu district of Nagano

Prefecture. As a consequence, piston ring maker Riken's factory located near the epicenter of the quake had to close down temporarily. This in turn meant that Toyota had to suspend production in all its domestic plants because the fine-tuning of its just-in-time delivery system simply did not allow for such an eventuality. As a matter of fact, Riken accounts for 50 percent of Japan's piston ring market and all 12 domestic vehicle manufacturers were forced into temporary stoppages. This is a world which the specialist LME has done much to create, and on which the large multinational has come to rely. And the solution to guard against being caught out again is not so far coming from the customer's side. Apparently neither Toyota nor any of the other auto makers are about to internalize the whims of the Almighty by bringing the production of piston rings in-house. On the contrary, Riken has voiced its intention to set up storage facilities close to its clients' factories and to disperse production, including to its plants in Michigan in the United States and Wuhan in China (Nikkei, 2007a).

***

In not a few cases the implications are worldwide: American and European multinationals may well be beholden to the inputs of the Japanese specialist manufacturer. Moreover, this is even more so for the neighboring East Asian economies. As noted above, time and again Japanese specialist manufacturers have actually been there peering into the cradle of the newborn technological advances in these countries, their expertise and output ready at hand to accelerate the process and to ensure themselves a hefty take from the value added into the bargain. From the outset they have been an integral and indispensable part of East Asian economic development. Not a little of the consequent production incorporating the inputs of Japanese specialist manufacturers has traveled the world. That having been said, many of these specialists are not so large and so can also be designated as Japanese leading *medium-sized* enterprises. How they came about in their national environment and what they face in the globalizing world comprise the subject matter of this book.

The sequence

A long-practiced art in depicting the Japanese industrial setup is to refer to it as a dual economy. The most fervent exponents of this approach are those who are intent on showing how much the economy is dominated by its larger conglomerates, frequently dubbed *keiretsu*, which are the most effective, innovative and talented in the land. But there is in addition a substantial army of scholars and commentators touting the cause of smaller firms, or SMEs, either commiserating with them for their disfavored status or exalting their role as fountainheads of structural renewal. There again, there are the critics who see SMEs as the unimaginative burden pulling at the national purse strings. What they all have in common is the conception of a chasm of varying degrees of depth and verticality dividing the economy in two. At worst this is not true and at best not helpful. The Japanese economy is far better seen as a spectrum of activity and actors, and moreover a spectrum which has seen the middle range of its coloring fill out especially over the past 50 years. This is the message as seen through the eyes of scholar Nakamura Hideichiro starting with his studies in the 1960s and as briefly recounted in Chapter 2. He detected what he called *chuken kigyo*, the leading medium-sized enterprises, or LMEs, which became the specialist manufacturers discussed in these pages.

Such companies are not peculiar to Japan. They have a universal provenance going back in some cases centuries, as Chapter 3 explains, although they have particularly flourished as the early industrial period has elaborated over the past 200 years into the era of globalization that we are now in. Universal though they may be, however, even now in their gestation and initial growth they still bear the hallmarks of their distinctive national origins, the institutions and attitudes surrounding their early existence influencing the ways they interact with that environment and how they tackle the world. The American, German and Japanese specialist firm looks outward through nationally prescribed glasses even as it devises individualistic solutions to tackling the world beyond, presaged on its unique expertise. There are similarities and dissimilarities then, and most of this book is devoted to the latter as it attempts to describe and analyze the particular destiny of the *Japanese* specialist manufacturer. Enough has been said, though, to suggest that in following this course it is worth bringing along a toolbox of references as to what such an idealized specialist could represent, and hence the pit stop in Chapter 4 to take on board the stylized LME. Most prominent in that toolbox, we find, are a hardnosed commitment to core competence, an engagement with change which renders the LME simultaneously the embodiment and agent thereof, and a mode of operation finely tuned to accommodate both constructive coordination and individual input. In its approach this idealized specialist expresses its ambition which has now embraced globalization.

Reverting to Japan, in pre-industrial times that country, like comparable Western societies at the time with their relatively sophisticated organizational structures, was hierarchical. Where it arguably differed to a degree was in the intricacy of the gradation from top to bottom. This gradation is symbolized in Chapter 5 by the elite, central 'samurai' at one end and the parochial, commoner 'artisan' at the other. Having been introduced as historical realities of 300 years ago, these concepts are subsequently employed throughout the rest of the book to epitomize the convergence and tension between different elements of Japanese society which have contributed to the shaping of its industrial structure and the firms functioning within it. Hence, from the outset of the Meiji era and externally enforced modernization starting in 1868 the samurai is striving to come to terms with the big picture on a national scale while the local artisan is reverse engineering and reconstituting the minutiae of the industrial age. Then Chapter 6 sees the process evolving from the two ends up to around 1930 as the samurai comes to interpret his role as orchestrator of a national industrial structure, in so doing delving deeper and deeper into the realm and the workings of the smaller firm. The artisan, meanwhile, having honed his modern trade is learning to link up more effectively with other firms, both large and small, both supplier and client. The exigencies of a war setting through the 1930s and into the 1940s then make this dual trend more explicit, as outlined in Chapter 7. The samurai, in seeking control of production, had to be selective. The artisan, in responding to an increasingly demanding environment, became more specialist.

In discussing the evolution of Japan's economy from 1945 to 1970, Chapter 8 pinpoints three essential motivating forces. First was the primacy of economic growth, bolstered by the newfound supremacy of the economic ministries. Second was the industrial planning formulated by these ministries to accelerate such growth. Their planning prioritized certain industries and designated certain firms for preferential treatment. This was a rush job to catch up with the more advanced countries of the West. It encouraged further specialization, which in Japan was realized more through the channels of specialist manufacturers than internalization by large corporations. Third, and likewise demanding urgent attention, was the unerring pressure exerted by international institutions like the International Monetary Fund for the liberalization of the Japanese economy and its opening up to the outside world. By the beginning of the 1970s, the LME was a distinctive and established element of the economy, just at the time, in fact, when new challenges were afoot in the form of new technologies and the evolution of software. This called for a new type of LME. It also engendered, as Chapter 9 relates, a diversification of responses as the international started to impinge more forcefully on the national stage.

Some of these responses had a non-Japanese flavor about them, but some remained characteristically Japanese, notably the commitment to closely defined core competences. This is then given further amplification in Chapter 10 which explains another determinant of Japan's development which was essentially home-grown and of long duration. This was *monozukuri*, or the national zeal for making things. Minute attention to detail, so characteristic of the Japanese, was a natural inducement to specialization. However, while the term *monozukuri* started out as a simple expression of the artisan's commitment as a craftsman, it has now been adopted by the samurai to encompass the whole gamut of business activity from product conception to after-sales services.

Chapter 11 then takes up globalization and the position of the Japanese specialist manufacturer within that process. Globalization, it notes, began to get into its stride in the 1980s. Its origins are to be traced to, among other factors, increased worldwide trade, and foreign direct investment. The Japanese had their own take on internationalization and were keen to promote their own institutions and practices. Although these institutions were inevitably diluted as they intermingled with the globalizing environment, the Japanese specialist LME still managed to foist its distinctive national identity on events, boosted not least by its own core competences based on its capacity for *monozukuri*. Assisted in its advances overseas by national institutions, it was also possessed of incentives of its own brought on by foreign demand and limitations like mounting costs becoming apparent on the domestic scene. What this entailed in detail is then given some airing with the introduction of a population of 429 Japanese specialist manufacturers in Chapter 12. These LMEs are on average only a fraction the size of a giant like Toyota and yet individually and as a group have managed to attain worldwide reach of sorts, establishing a relatively strong presence in North America and Western Europe, and an especially strong manufacturing presence in East Asia. The latter being so, the Japanese LME is now an indispensable participant in East Asian industrial development. At the same time as this LME ventures outward, however, globalization also impinges on its home territory, forcing it to make the necessary adjustments. This is the subject of Chapter 13. Cases of merger and acquisition usually amicable and intra-industry to cope with cheaper imports, for instance, are on the increase. More ominous, on the other hand, is the emerging menace of hostile takeover heralded by the appearance of investment funds intent on implanting alien notions of shareholder precedence. So far this is being kept at bay. But the possibility of Japanese specialist manufacturers being taken over and dissected is not being dismissed lightheartedly. Nor is the rise of China. Again, so far the Japanese LME has in the main been able to accommodate and indeed benefit from China's industrialization as it has with other East Asian countries. But if any country is capable of breaking the mold it is China.

2
The Dawning of a Concept

Nakamura Hideichiro designates the LME

At the beginning of the 1960s, the Japanese economy was champing at the bit on the premonition of imminent explosive growth. In fact, incumbent prime minister Ikeda Hayato felt confident enough of such prospects to unveil and launch his Cabinet's income-doubling plan for the nation. Preparations were also in the works for Japan's post-war symbolic reinstatement within the community of nations through its holding of the 1964 Olympic Games. Japan was once more a fully viable, functioning socioeconomic phenomenon whose structural makeup merited detailed perusal. Not untypical of those surveying the landscape was a young scholar called Nakamura Hideichiro. He was researching for a book on the perceived problems of what in Japanese are referred to as *chusho kigyo*, and in English as small and medium-sized enterprises, or SMEs.

In writing such a book, Nakamura was in keeping with the times. He was, needless to say, conversant with the route the economy had taken over the one and a half decades since the end of the Second World War and the structure it was purported to have assumed. The received interpretation went that in effecting Japan's rehabilitation the emphasis dictated by the political, bureaucratic and business elites had been on prioritizing what were deemed to be strategic industries. These were coal mining, shipbuilding and steel making, subsequently joined by the machinery and electronics industries. All of these industries were thought to be the province of firms which were large by Japanese standards, supported and complemented by large financial and trading institutions in a world which at the time unequivocally embraced the merits of the corporate leviathan. Within the hierarchical business structure which was – at least from the elitist samurai viewpoint – the essence of Japan, there was a distinct divide between the large, efficient and nationally committed as against the small, haphazard and parochially immersed. In a word, Japan's destiny was seen to rest on a dual economy where the large enterprises tackled the real world, while the small were tolerated as mundane suppliers and recession-easing buffers, although also as socially indispensable absorbers of excess labor and useful preliminary exporters until the economy got off the ground.

There were historically ingrained precedents for this attitude, but reasons can also be found in the bumpy road to rehabilitation that Japan experienced in that first decade or so. During that time, funding was constrained, technology was backward, materials were scarce, and jobs for many were hard to come by and retain. With preference afforded to the large and influential in the interests of reviving the corporate establishment, SMEs often walked the tightrope of under-financing and materials shortages, their predicament only temporarily alleviated, initially, by abundant subcontracting orders received in the procurement boom brought on by the Korean War from 1951 to 1953. But this then came to an abrupt halt; order volumes to SMEs plummeted, prices fell and payments from big and powerful customers were delayed. SMEs had to wait for the export-led economic recovery starting 1955 which extended ultimately, hiccups notwithstanding, to the high-growth era of the 1960s to enjoy relative stability and less demeaning exploitation.

So an admixture of reality and prejudice contrived to paint one image of the Japanese economy and the SME component within it. That is, the dual economy of the large and the small and the implications inherent therein. However, such an image was not universally accepted, and Nakamura was to become even more of a detractor than when he started as an unanticipated outcome of his research. Stemming from the postwar Occupation, when the Americans endeavored to invest the Japanese business environment with more democratic principles, the left of the country's ideological spectrum had voiced demands for a more level playing field where smaller firms would enjoy access to financial services and business opportunities on more equal terms with their larger counterparts.

Herein lay the seeds of what in a couple of decades or so would blossom into the notion of SME as network participant rather than subservient supplier, and as an essential, complementary actor in sustaining and elaborating Japan's industrial economy. Moreover, ideological advocacy aside, emergent economic forces were already, by the middle of the 1950s, increasingly insinuating new potential for the smaller firm with initiative and drive. The number of SMEs in manufacturing was on the rise, despite business volatility and unpredictability, and there was an intensification of the pattern of subcontracting relations. Growth, it seems, by the end of the 1950s, was becoming endemic. To accommodate expanding markets both at home and abroad, the large firms actively progressed with plant and equipment investment; but to achieve their ends in such conditions they were also obliged to resort to their subcontracting relations with SMEs to make good their own resource limitations. Having been prevalent notably in the textile industry in the prewar period, this subcontracting system now extended its hold more thoroughly over the likes of machinery manufacturing and metal processing. Such a turn of events, admittedly, was not all good for SMEs. It meant that they tended to be sucked into the maw of big corporation dominance, beneficiaries of upturns but victims of downturns. But on the other hand it did show, given the structure of the Japanese economy as it assumed a mode of accelerating advancement, that SMEs as a genre were an integral and indispensable part of the country's renewed march of progress (Teraoka, 1997). For one, by the mid-1950s, with their burgeoning production of light machinery, toys, footwear, apparel, bicycles and what have you, SMEs were displaying the competitive side of Japan and, in so doing, accounting for over half its exports.

By systematizing and synthesizing SME theory, Nakamura aimed to show that the SME was encompassed within, rather than peripheral to, the scope of the Japanese economy. In other words, he started out as one of those detractors out to prove that SMEs were already playing a significant role in the national economy. As this endeavor progressed, however, he noted that there were a surprising number of companies which had grown to the point that they exceeded the framework of the SME and that, in a strictly dual conception of the economy, there was no place to position them. Having conducted further examination, he reached the conclusion that they were no longer SMEs, but instead constituted within the Japanese economic structure a third type of company qualitatively different from both the general SME and the large corporation. They therefore merited a distinct designation: *chuken kigyo*, leading medium-sized enterprise, or LME.

A creature of its times ...

Although, as will become apparent, many of these LMEs had their origins in prewar times and, in some cases, going back to the Meiji and even Tokugawa eras, they epitomized for Nakamura the spirit of that era of the early 1960s, an era not only characterized by fast growth but also a progressively consuming commitment to quality. And quality, more than anything else, impressed Nakamura as being intrinsic to the LME. Quality permeated its decision-making, was manifested in its product development, manufacturing technology and marketing creativity, and realized a superior capacity for raising capital, which in turn allowed for the installation of the latest machinery and equipment and the hiring of individuals a cut above the average (Nakamura, 1990:2–3).

These LMEs increased in number in the high-growth era in tandem with the upgrading of the nation's industrial structure as it shifted in emphasis from light to heavy industries and enjoyed rapid market expansion. This process caused a flowering of innovation galvanized by imported technologies, transformed the domestic market structure, and advanced the specialization of production, in so doing intensifying the division of labor within the economy. LMEs were beneficiaries of specialization and mounting expertise, having been contributing instigators of innovation and structural transformation. This was not least because they were heavily concentrated at the future end of the industrialization time perspective. In the early 1960s, Japan's products could be broadly posited in three categories: products which were already internationally competitive such as textiles, steel and light machinery; those which lacked international competitiveness as most clearly illustrated by agricultural products, especially rice which was priced 30–40 percent above the price of rice produced in Southeast Asia; and those which, with economic development and upgrading of the industrial structure, continued to expand, but were possibly not as yet internationally competitive, such as electronics, auto parts and machine tools (Zhu, 1993:47). Needless to say, this last category – machine tools – is where many an LME was to be found and as such will preoccupy us anon. Suffice to say at this juncture that, having weathered the uncertainties of the 1950s, machine tool companies like Mori Seiki and Makino Milling Machine, for instance, single-mindedly pressed forward in their core competence to start flourishing in the 1960s and rise in stature up to the present day.

One major stimulus for the elevation of certain smaller firms to LME status was the establishment among the large assembly plant industries of full-scale mass production commencing in the mid-1950s. Because, as will be elucidated in more detail below, there has been a tradition whereby Japanese corporations tend to be smaller, more highly specialized and more reliant on outsourcing than their Western homologues, this mass production by assemblers called for the development of a new type of SME capable of effecting a corresponding mass production of parts, components and tools. Failure to make the grade in this respect at that stage in the evolution of the economy was in fact tantamount to forfeiting the chance to attain LME colors. Under these circumstances, new entrepreneurs emerged possessed of a different temperament and outlook than the run-of-the-mill SME operator. Eschewing the tunnel vision of mere personal gain, these new entrepreneurs grasped the potential for long-term commitment that progressive division of labor presaged and devoted their attention and energy to building up their enterprises so that they were capable of not only meeting face-on the fast-moving targets of prevailing demand, but also constructively preparing the groundwork for what lay ahead.

These entrepreneurs were at the leading edge of managerial reform of the smaller company, while they also challenged generally accepted ideas in the industries of which they were a part. The strategies they adopted in lifting their companies out of humdrum SME orthodoxy to LME ranking typically entailed three elements. The first was to select products which were original and had growth potential. Second, through the introduction of various innovations, the intention was to achieve product mass production and quality upgrading so as to develop the market and ensure scale merits as quickly as possible. This was with the aim of establishing competitive advantage over other SMEs as well as big companies who were or potentially could be rivals. Third was the aggressive investment in equipment and the introduction of new processing technology, to be fused with a committed workforce whose individual capabilities were recognized in an atmosphere of mutual respect. Important here was not only bringing in top-class machinery with drastically improved processing, but also internalizing equipment so as to develop production knowhow, both of which were decisive factors for cost competitiveness.

... keeping pace with events ...

This is what Nakamura detected in the early 1960s. He saw no reason to recant this stance when he revised and updated his observations for a new book on the subject published in 1990. Quite the contrary, in his view the orbit and pertinence of LMEs had been further augmented, as they embraced knowledge-intensive industries in the 1970s and wider diversity of production within the scope of core competences in the 1980s. While his main preoccupation was with manufacturing LMEs, he also extended the term to cover distribution, wholesale, retail and other services. Because the LME is first and foremost a *type* of business enterprise, he was less concerned with size *per se* as denoted by capitalization and number of employees, although of necessity the many examples he cites occupy a capitalization range of ¥40 million to ¥1 billion and accommodate 200 to 2,000 employees. What he is more interested in is the dynamics. LMEs epitomize movement, both in the temporal and spatial environment in which they function and in their accommodation to that environment. Having elevated themselves from the SME fold, they may rise still further to become large corporations, or sink into oblivion for lack of agility in adjusting to change, or – most importantly for this study – remain where they are out of choice as committed and proactive LMEs.

Encapsulating the cut and thrust of this dynamic most dramatically was the crucial phase from the mid-1960s to the mid-1970s, which commenced in exhilarating high growth and ended in the recession brought on – albeit temporarily – by the first oil crisis. It was during this phase that Japan's relative isolationist protectionism was forced to give way to a first taste of the modern variety of internationalism. In 1965 high growth in Japan was largely fenced in from undesirable foreign competition; by 1975 labor-intensive industries were in rapid retreat as the inevitability of the need for upgrading to knowledge- and capital-intensive became glaringly apparent. Reasons for this transformation of circumstances included, first, Japan's commitment to the General Agreement on Tariffs and Trade (GATT; the forerunner of the World Trade Organization), the International Monetary Fund (IMF) and the Organization for Economic Cooperation and Development (OECD), which obliged it to liberalize and open up its commercial and financial institutions. Second, on August 15, 1971 American president Richard Nixon delivered a double blow: he announced a ten percent surcharge on U.S. imports, while at the same time taking the dollar off the gold standard and inaugurating the floating exchange rate system. The surcharge seriously threatened the survival of Japanese makers of toys and shoes. The immediate effect on the yen, which had been pegged to the dollar at 360 for years, was for it to appreciate markedly, not only against the dollar but also against other key currencies at the time such as the pound sterling. Finally, there was the oil shock itself, putting an end to cheap energy, and thereby adding further stimulus to the quest for knowledge-intensive solutions.

Japanese LMEs walked the boards of this drama. Some fell through the cracks. Perhaps, for example, other than achieving volume production, those who could not make the grade remained undistinguished from

SMEs in not progressing with technological development and marketing skills. In other cases, having pioneered one successful line they lacked the ability to follow up with the next effective product. In still others, despite having the technology and production process well in hand, they could not cope with the managerial and organizational problems attendant upon growth. But overall, this decade saw a healthy augmentation of the LME ranks. These were the LMEs who applied themselves constructively to developing production capabilities, deepened their specializations, erected their own entry barriers, and ensured themselves high market shares. The background to this was change in what and how companies with LME pretensions were called upon to deliver. Starting in the latter half of the 1950s they could get away with manufacturing parts using designated mass production machines, and assemble components employing abundant cheap manpower. In the 1960s by contrast, with progressive wage hikes and the absorption of excess labor in a fast-expanding economy, the ambitious LME looked to upgrading the effectiveness of its workforce and combining this with more advanced automation. Knowledge and capital were beginning to supersede brawn.

Spicing this trend was the growing complexity and diversity of demand which could now be met by the progressive upgrading of production control technologies. This led to advances in small lot production of a wide variety of categories, often complementary to continued mass production. Hence did Kato Spring Works (recently renamed Advanex) come to develop many different types of spring which, according to materials used and configuration required, ran the gamut from being almost completely handmade, to semi- or fully-automatically manufactured, in lots ranging from a mere 100 to 100 million (Nakamura, 1990:384). In addition, now increasingly confirmed in the depth and intricacy of their chosen specializations, as the 1960s gave way to the 1970s, many LMEs were starting to spread their know-how potential over wider catchment areas, taking on new industries and customers in the process. To give but one example, having started upon its establishment in 1954 by supplying toy manufacturers, miniature electronic motor maker Mabuchi went on to furnish the audio equipment and timepiece industries in the 1960s and then the producers of cassette tape recorders and automatic vending machines in the 1970s. Moreover, surveys undertaken at the time began to imply that there was not much difference, if any, in sales rate per employee between LMEs and large corporations, suggesting that the more advanced the level of industrialization, the more scope there was for specialist firms. Herein was the terrain in which LMEs could exist and prosper.

... while internationalizing and rising in status

Another distinguishing feature beginning to set LMEs apart from the SME crowd by the advent of the 1970s was internationalization. Some, like Mabuchi – which had put down roots in Hong Kong as early as 1964 – were already making headway with foreign direct investment, especially in the their East Asian neighbors and the United States. However, the main form of internationalization at this stage was exporting. Assisted by the increasingly free trade regime, LMEs were becoming international specialist manufacturers. As noted earlier it was *SMEs* that were at the forefront of Japan's exports in the mid-1950s, responsible for some 58 percent in 1957. But a decade later, the ratio for SMEs in the January to September stretch for 1967 had fallen to 43.3 percent according to the government's SME White Paper at the time (Nakamura, 1990:390). There were three reasons for this, the first being that exports of traditional, labor-intensive products which had been to then an SME forte were in decline, both absolutely and relatively. Second, large corporations were by then in the full swing of expanding production and progressively capable of turning their attention to overseas possibilities having, in many cases, used the domestic market as a springboard. Third was the growth in exports by a new type of specialist, many of whom could by then be designated as LMEs, having outgrown and outperformed SME status and now capable of handling exports on a considerably larger scale than typical SMEs.

These exports could be end-user items like the unique writing instruments emanating from Pentel, but as likely as not they were highly sophisticated intermediate adjuncts to the assembled product: poly-variable condensers for transistor radios from Mitsumi Electric, bicycle gears from Shimano. What these LMEs had succeeded in doing was, with the domestic market as base, raising their technological level, establishing mass production systems, upgrading production quality, and reducing costs, thereby gaining international competitiveness. In many cases their products were original and could fetch healthy profits on the world market. Such LMEs were themselves beneficiaries of advances in related industries allowing for upgraded materials and machinery and equipment, while further refinements in the division of labor enhanced the quality of the parts and components at their disposal.

Funding was also more readily available to *bona fide* LMEs as they found their way up the banking hierarchy of preferred clients. In addition, although even nowadays by no means all LMEs are listed, one tactic open to them at the time was to go public by getting registered on the Second Section of the Tokyo Stock Exchange, for example, as a step in the direction towards the more prestigious First Section. The Second Section was inaugurated in 1961 and by the end of 1963 boasted a membership of 583 companies, indicative of the fast-growing demand for equity funding of a fair number of medium-sized firms as the nation's economy geared up to full-fledged take-off.

The watershed

In all this Nakamura saw a watershed dividing the 1960s from the 1970s in the evolution of LMEs, the pervasive background to which was the irrepressible tide of internationalization. Japanese exports could no longer be expanded at will because of the demands from advanced countries, particularly the United States, for currency adjustment and voluntary restraint regarding exports; smaller firms in the labor-intensive industries were feeling the pinch as competition intensified with developing countries, notably Taiwan, South Korea, Hong Kong and Singapore – subsequently to be dubbed the Asian NIEs – as they started to learn the ropes of industrial production and overseas marketing. Simply put, Japan was under increasing pressure to readjust its comparative advantages, a restructuring which looked more and more to taking on an internationally-oriented division of labor. This was the challenge facing Japanese LMEs who now had to redefine their course if they wished to keep pace with the pack.

In this light the 1960s can be seen as the first phase in a rite of passage. It was then that many a putative LME selected its product line thereby instilling its core competence, although this selection could be manufacturing something for the home market which had its origins outside Japan. This LME also developed mass production and mass marketing techniques, upgraded and internalized top-quality machinery and equipment, and perhaps made progress in the separation of ownership and management, albeit more in spirit and awareness rather than actually offloading founder and family. Typically the LME commenced business with a local clientele, progressively building up its market share until in a position to break out of the regional and onto the national stage, and thereafter putting out feelers abroad. High export ratios could be achieved because Japanese wages were still low, assuming such exports were headed for the advanced world, which many of them were.

However, this approach was no longer state-of-the-art as the 1970s unfolded. To survive the LME had to show its mettle through entrepreneurial innovation. This meant originality and differentiation. Arriving at these goals, in turn, called for reinforced R&D and design, divestment of the obsolescent, systematization of markets, outsourcing of non-core elements, professionalization of management, and – inevitability – attention addressed to the scheme of things in the world at large, to what in a few years time would come to be called globalization. The LMEs which thrived in this new phase were those which brought out innovative products and revolutionized manufacturing processes, often by such means gaining international recognition for their efforts. Nihon Miniature Bearing, for example, did this by solving the problems of miniature bearing mass production, thereby ensuring its competitive advantage on an international scale. Conscious also of the need for an upbeat image in a wider world, this company also changed its name in 1981 to what it is now, Minebea.

Acting as they did, moreover, these LMEs were also manifesting an ever greater degree of autonomy. Even though they had progressed remarkably in the 1960s, they were still essentially playing a supplementary role in an industrial structure geared to the heavy and chemical industries. The emerging LME character of the 1970s, on the other hand, in its transformation to knowledge- and technology-intensive industries, began to assume a self-motivated *raison d'être*. No more beholden – if they ever had been – to one or just a few key clients, LMEs began to spread their wings in networking arrangements made up of suppliers, alliance partners and customers rather than being captive to restrictive pyramidal hierarchies. In so doing LMEs reemphasized their specific identity as specialized medium-sized firms, although now raising their flag up a notch higher. That is to say, having in the 1960s rejected the short-termism which typified many an SME owner-manager, by the 1970s LME leadership was presenting a challenge to the rigidly controlled large corporation slow to adapt to industrial structural change.

The LME established

By the 1970s, in fact, the term *chuken kigyo* – LME – was perfectly familiar in Japan. LMEs were by then significant players in the economy. A survey by the Japan Fair Trade Commission in 1971, for example, claimed that there were by then some 10,000 independent LMEs and saw them accounting for 20 percent of the economy's capital and profit ratios. During that decade, moreover, their number was to more than double. Another survey conducted in 1971 by the Japan Long-Term Credit Bank reported that 22.6 percent of LMEs were guaranteed markets totally different from those of large corporations, 37.8 percent competed with large corporations effectively with their own technologies and sales routes, and 35.5 percent had large corporations as clients and seemed to be enjoying an unchanging situation of mutual dependence. From this it concluded that no less than 95.9 percent of these LMEs were ensuring their own markets. What is more, of the LMEs surveyed, only a mere 1.8 percent said they were absolutely set on becoming large corporations (Nakamura, 1990:427).

Then, throughout the 1980s, a growing number of LMEs were listed on the Second Section of the stock exchange, not a few of them thereafter elevating themselves to the First Section, which often meant that they went beyond the capitalization ceiling of ¥1 billion (Nakamura, 1990: 424–7). A lot of this, of course, depends on definition as to what an LME is reckoned to be, and this will be discussed in more detail below. Suffice here to remark that this book does not include such a large number of firms within the LME category as the abovementioned survey does; it is more selective. But two comments are worth making. First, it is interesting to note that many companies were happy to designate themselves as LMEs with the self-limitations and bounded scope that this implies; depth in expertise and technological know-how took precedence over broad-based diversification and expansion. Second, the LME can be seen as a flexible being keeping time with the rhythms of the economy as it evolves. Hence, although LMEs can now be found which are capitalized on a par with companies deemed as large, they have nevertheless retained their LME *character*. This is why, even at the turn of the millennium, at least one Japanese commentator could still refer to Minebea and Mabuchi Motor as LMEs, for example (Tsuchiya, 1999).

LMEs, then, constitute an established feature in the Japanese economic landscape. By business gurus and marketing specialists they are written up as the stars which ordinary SMEs, and even large companies, should look to emulate. The media regard them as a breed apart and banks design financial packages for them, although – unlike in the case of SMEs – there is no government agency tending to their specific needs and little official recognition of their singular identity. It goes without saying that the term *chuken kigyo*, thus coined, is unique to Japan and that the choice of expression reflects the evolution of that nation's distinctive corporate structure. Indeed, companies identified as answering to the description of LME are being employed in this book to illustrate a particular contribution by Japan to globalization. However, whereas a country's institutions can in their finer details claim uniqueness, regarded from a higher elevation the contours of similarity with the institutions of other countries become apparent. Taking this broader perspective, this type of company is in no way Japan's sole preserve. LMEs, albeit labeled differently, have originated in other economies – in the United States, Germany, France and more recently in Hong Kong, Singapore and Taiwan to name just a few. This being so, far from being incidental to the debate, LME lookalikes from around the world, as well as what they do and how they do it, are indispensable references never absent from the screen. Just like the Japanese LMEs discussed here, they too are dedicated specialists initially nurtured in familiar home territory and now increasingly absorbed, actively and reactively, in a process of change called globalization.

Summary

When Nakamura Hideichiro started his research on a book about SMEs at the beginning of the 1960s he subscribed to the received opinion that Japan had a dual economic structure comprising large corporations on the one hand and small firms on the other. His task, as he saw it, was to characterize SMEs in a more positive light than had hitherto been the rule, given the strengths they were revealing by the end of the 1950s as effective subcontractors and exporters. As his research progressed, however, Nakamura realized that there was a third kind of company emerging which was distinguished by its quality as specialization increased, as well as by its ability to handle mass production. He called such companies *chuken kigyo*, which is rendered here as leading medium-sized enterprises, or LMEs.

Nakamura saw no reason to revise his opinion when a rewritten version came out in 1990. In fact, what with the development of knowledge-intensive industries and evolving diversity they were more in evidence. Moreover, it was their dynamic which set them apart rather than their size *per se*. During the 1960s shrinking unemployment and mounting international pressures for Japan to open its markets challenged LME contenders to upgrade equipment and personnel. They responded with greater efficiency and elaboration of production lines in the quest for a broader customer base. In addition, they raised their status by getting listed on the stock exchange and learned to translate their domestic successes into export potential. Having established their credentials in the 1960s, LMEs were a recognized force by the 1970s. But with high growth coming to an end and shocks overturning the assumed order of things, from here on knowledge and innovation were to be the keys to survival for the LME which is prevalent in Japan but not unique to it; homologues exist in other countries around the world.

3
The Universal LME

An American and a German

In the literature of the firm there has been a tendency towards polarization in the discussion. Typically, in the early writing the large corporation was considered the archetype to which all lesser firms should ideally aspire in their conduct of business, while from the 1980s especially, this assumption has been met with a counterthrust pointing to the differences, merits and distinctive functioning of the smaller enterprise. Little attention has been directed towards the medium-sized firm as an entity in its own right. All too often it has been subsumed under the rubric of the large or arrogated to the cause of the small. In the first case, implicitly at least, the impression is given that a company with a workforce of, say, 600 somehow has the capacity to act like General Motors with its hundreds of thousands. In the second, inclusion of relatively heavy-weight medium-sized enterprises in an analysis of SME activity can raise the performance mean, thereby somewhat distorting the picture as to what the average small firm – if indeed there can be such a thing – is capable of.

However, the presence of medium-sized firms has not been completely ignored. Two accounts of particular interest here are *To Flourish Among Giants: Creative Management for Mid-Sized Firms* by Robert Lawrence Kuhn, which came out in the mid-1980s, and *Hidden Champions: Lessons from 500 of the World's Best Unknown Companies* by Hermann Simon, published in the mid-1990s. Special attention is drawn to these books and the companies cited in them for a number of reasons. To start with, their titles exude the same positive ring inherent in the term *chuken kigyo*, or as I have rendered it in English, *leading* medium-sized enterprise. Second, like this book and therefore complementing it, neither of their authors aspires to the pinnacles of scientific robustness. Instead, they are content to explore the character of their elected genre of enterprise and the domain in which it operates in a relatively anecdotal fashion in order to engage the reader's imagination as much as to present an unconventional angle on what goes on in the real business world.

Third, together the two books constitute a contrast which can in turn be employed to contour the particular nature of the Japanese LME in its own home territory and in the international arena. Kuhn is talking exclusively about American medium-sized enterprises, while Simon concentrates largely – although not entirely – on German counterparts. The contrast, therefore, is between companies, on the one hand nurtured in the cradle of quintessential Anglo-Saxon liberalism and, on the other, brought up in the heart of a productivist culture. Or, as Ronald Dore (2000) would have it, it is a contrast between firms raised on stock market capitalism and those raised on welfare capitalism. While the American paradigm is to a fair measure antithetical to, albeit at the same time presenting a challenge to, the Japanese version, the German framework has marked similarities with the Japanese.

The fourth and final point is that the two books offer a time perspective which renders further relief to the above contrast. Even as he wrote, Kuhn noted that the position of medium-sized companies in America was weakening, occasioning him to predict that they were in for further decline *vis-à-vis* that nation's industrial structure as a whole. And this is indeed what has arguably happened, in the more traditional manufacturing sectors at least, during the intervening years; in an economy driven by financial imperatives and shareholder value, and given the need to respond to globalizing pressures, if they have not expanded themselves, not a few successful medium-sized companies have been outstripped, sidelined or absorbed by more powerful or more adept players. Alternatively, coming a decade later, Simon is decidedly upbeat about the role, and indeed the survival, of the same kind of core-competence, specialist enterprise, not only in its mainly German home environment but also, very much so, on the international scene. In fact, going international for these 'hidden champions' was seen by him to be an essential and inevitable market-enhancing strategy allowing them to fulfill the potential for their products, being as they were often subject to highly specific and restricted demand on the purely domestic front.

But here, too, the situation has evolved since Simon put pen to paper. The same seemingly irrepressible ascendancy of financial institutions and the logic presaging their actions, adumbrated by Kuhn, is nibbling at the foundations of the productivist ethos. German companies, too, are witness to the exigencies of change; increasingly self-interrogation must dwell on the issue of 'to buy' or 'be bought', driving a wedge between traditional values deeply beholden to a preoccupation with technological prowess and a future path in all probability entangled in as yet less familiar strategic juggling. Japanese LMEs are in the same boat, if anything more so because of their non-Western origins, the national institutional framework in which they are embedded, and the particular structure the country's economy assumed, especially in the post-Second World War period.

So later on we shall be taking time out to look at a number of aspects which can be regarded as characteristic of American and German counterparts the better to understand what the Japanese LME is, how it operates and where it is heading. However, highly distinctive in their respective ways though they may be, these three countries are not the only breeding grounds for the LME type of company, particularly among the community of advanced industrialized nations. In pursuing our analysis of the LME, therefore, we can cast our net over a broader catchment area, if for nothing else but to show that the LME has some claim to universality. This being the case, we shall start by tracing the historical evolution in which a number of companies which can be categorized as LMEs from countries other than Japan, the United States and Germany have developed. The chapter then culminates with an exploration of how a select group of non-Japanese LMEs have set about internationalizing and globalizing. The observations thus gleaned, coupled with the depiction of the stylized LME in the next chapter, constitute a reference backdrop for when studying the evolution of the Japanese LME as the book progresses thereafter.

The LME in history

The subject matter of this book comprises companies which can be currently classifiable as LMEs, or which have been LMEs until recently, having passed through a formative LME apprenticeship. They can have started as apothecaries or blacksmiths centuries past, or leapt on the ultra-modern venture capitalism bandwagon a decade or so ago, or anything in between. The point is that the LME phenomenon has its place in historical evolution and examples can be abstracted from a broad spectrum. These LMEs can also be posited in discernible stages of this evolution, as five of them have been in Figure 3.1, one each to represent what are referred to here as perennial, industrialization, postwar, state-of-the-art, and emerging-economy LMEs.

Figure 3.1   LMEs Evolving

Company	Founded	Country	Product	Historical Status
Codorniu	1551	Spain	Wine	Perennial
Lego	1932	Denmark	Toy bricks	Industrialization
Brembo	1960	Italy	Auto parts	Postwar
Gilat	1987	Israel	Satellites	State-of-the-art
Sunplus Technology	1990	Taiwan	Semiconductors	Emerging economy

Perennial LMEs

Perennial LMEs are deemed here to be companies which can trace their origins to before the take-off of the industrial revolution, whenever that may have been in the country where they initially commenced business. Consequently, they are associated with traditional industries such as alcoholic beverages, textiles, ceramics, writing instruments, medicines, and metals. Prolonged family ownership is one of the distinctive hallmarks of perennial LMEs, as it is with early starters in the industrialization phase. In France, for example, even now the medium-sized family-based company, referred to as *la moyenne entreprise patrimoniale*, retains a significant role in that country's economy and has been described engagingly as "a curious animal which merits a special place in the very varied zoology of private enterprises" (Gélinier, 1996:135), a sentiment resonant of Nakamura's view of the specificity of the Japanese LME.

Codorniu of Spain answers perfectly to that format also. The first written record of its existence goes back to 1551 when a certain Jaime Codorniu was bequeathed cellar installations and utensils for cultivating vines. Only the owning family name has changed, for in 1659, when heiress Maria Anna de Codorniu plighted her troth to Miguel Raventos, it became Raventos. It has remained so to this day with the Raventos family as 100 percent owners (www.codorniu.es). Since 1998 the company, still Codorniu, has been chaired by Maria del Mar Raventos Chalbaud, while the managing director is Jordi Raventos Artes. Codorniu's transformation to accommodate the modern world commenced in the 1870s with its development of a sparkling wine, a path it nevertheless traveled in step with tradition, heralded by the company's elevation to the status of 'Purveyor of the Spanish Royal Household' in 1897. Upon celebrating its 450th anniversary in 2001, Codorniu was still a royal supplier.

Industrialization LMEs

These LMEs epitomize not only the essence of the Industrial Revolution and its first century or so of gestation but also a spirit of individualism and adventurism. They were the creation as often as not of one or two founders possessed of skills and techniques derived from the accruing knowledge of progressive industrialization. These founders plied the paths of new inventions, new technologies, and new materials, intent on 'making things' – an expression we shall return to again in discussing the Japanese LME later. Although not the preeminent drivers of change, industrialization LMEs nevertheless navigated the confluence of craftsmanship and scientific invention in pursuing the multiplying threads of opportunity as economic infrastructures were elaborated and markets of growing complexity emerged with the spread of accumulating affluence. Those established in the mid-19th century were small because business ventures were small at the time, so they preceded the enlarging consequences of organizational transformation that started to take place at the turn of the century. Conversely, those established at later stages were able to sidestep the issue in their elected, tightly-defined specializations as division of function became increasingly intricate. They presented themselves to serve the diversifying demands of both industry and end-user consumers with metal cans, bookbinding machines, sanitary fittings, industrial measuring instruments, security fences, coffee roasters and on and on.

It is important to constantly bear in mind that, as well as factory discipline and regimentation, the Industrial Revolution was the wellspring of a vast and intensifying elaboration of differentiation and opportunity. Manufacturers were presented with boundless scope for innovation; consumers with deepening pockets were increasingly able to make purchases tailored to their individual desires. This in turn became the elected pursuit of specialist future LMEs as they stamped their inimitable authenticity on terrains of their own making. Moreover, for infinitely more people individual desires were afforded expression in what they did when they were not at work or sleeping it off, in the greater time for leisure activities; they went cycling, read novels, and they looked to their children's enjoyment and fulfillment. This latter likewise invited specialization from which Lego took its cue.

To be precise, carpenter and founder Ole Kirk Christiansen did not take off with toy bricks. He is one of those instances where the crucial idea for success was revealed as his company was going about more mundane chores. That is to say, when he established his business in Billund, Denmark in 1932 it was for manufacturing stepladders, ironing boards and wooden toys. It was only two years later that, with a full complement of just six or seven employees, he dreamed up the toy brick. Subsequently, in the latter half of the 20th century, the company's durability, innovativeness and adaptation to changing demand – all essential prerequisites of a viable LME – are demonstrated by four moves. The first was when the warehouse for the wooden toys was destroyed by fire in 1960. Turning calamity to advantage, the decision was made to discontinue wood as the base material and concentrate exclusively on plastics, shifting first to cellulose acetate and then to ABS (acrylonitrile butadiene styrene). Second, company identity was reinforced in 1973, when a single new Lego logo replaced the various logos used until then. Third, in keeping with the emerging trend of linking toys with learning, in 1980 Lego introduced its Educational Products Department, afterwards renamed Lego Dacta. Finally, there is the company's ready embrace of new technology, as with the incorporation of a CD-ROM containing building instructions in the package in 1997. By just past the turn of the millennium Lego could boast that, over the previous 60 years worldwide sales of Lego bricks had exceeded 320 billion. It was the only European toy maker in the world's top ten, while remaining throughout an unlisted, wholly family-owned enterprise (www.lego.com).

However, the new millennium brought with it its own set of problems. Despite the pre-eminence of its core products Lego was revealing that an LME, even of its standing, was not impervious to the transforming business environment. Although in dealing with the situation it has again demonstrated true LME resilience and alacrity. There were two things wrong which, combined, were the main causes of a calamitous operating loss of some US$240 million and escalating debt in 2003. First was the problem of shrinking markets for high-end toys, and second was cavalier venturing into unknown territory way beyond the company's established core competences. The family reacted by bringing in an outsider as CEO. In the subsequent turnaround launched in March 2004, core products and core markets were reaffirmed; the employment roster was almost halved; some production was moved to factor-cheap Eastern Europe and Mexico; assets in America, South Korea and Australia were sold off. At the same time, the diversification strategy was refashioned to be more closely aligned with the established core competences (*The Economist*, 2006:82). The transforming global environment, the role of the outsider, and diversification are recurring themes throughout this LME narrative which follows.

Postwar LMEs

The Industrial Revolution multiplied the opportunities for trade regionally, nationally and then internationally. That is how industrialization LMEs established their pitch. But the road could be rocky; after the *belle époque* of expansive internationalization culminating in the late 19th century came the dark ages of protectionism and the dismantling of hitherto converging interests, which had the effect of arresting globalization in its tracks through to the Second World War. This was to change with the inauguration of the Bretton Woods system in 1944, directed at expanding trade under a fixed exchange rate regime and erecting the infrastructure of facilitating institutions in a new world order – the World Bank, the International Monetary Fund, the General Agreement on Tariffs and Trade (predecessor of the World Trade Organization). Subsequently, as post-war recovery materialized, confidence soared and the conviction took hold that the era of abundance had arrived, foremost in the advanced industrialized countries but increasingly elsewhere too. In the minds of many, innovation, growth and continuing prosperity had become economic staples. What is more, over the 35 years to the early 1970s, these staples were beginning to spread by way of start-up manufacturing ventures in new areas of the world, in Brazil and Mexico in Latin America, but most especially in East Asia.

These economies, then, were to a considerable degree the beneficiaries of mounting Western affluence presaged on growing technological and organizational sophistication. But in the West itself, notwithstanding the accelerating tempo and the compounding intricacy of change, traditional entrepreneurship still found plenty of space to express itself. Many a manufacturing LME in North America and Europe – not to mention chief emulator Japan – was started on a shoestring in the 1950s and 1960s, a trick largely relinquished to Silicon Valley come the 1980s. Brembo, the specialist in braking systems, brings together some of the distinctive strands of this era. On the one hand it presents the classic case of a modest family business, having been initiated in a small garage in Bergamo, Italy. But on the other, the scene is no longer the parochial hills enclosing the arduous life of the blacksmith-turning-machine maker in the early stages of industrialization.

Rather, it is wedded to the affluence and advanced technologies already in place. For with experience gained in mechanics and metallurgy, founder Emilio Bombassei commenced operations by doing one-off jobs for the likes of Alfa Romeo no less. By the mid-1960s the base of Brembo's business broadened considerably when the company became the first in Italy to manufacture brake discs for the aftermarket, this being followed by other braking system components. This was still with a workforce of only 28. Ten years later in 1975 it had 146 employees, with a turnover of 2.8 billion lira, and Brembo had gained leadership in Europe's brake disc aftermarket. In that year too it was entrusted with fitting the Ferrari Formula 1 car, clearing the path for it to soon rise to pre-eminence in braking systems for racing. This partnership with prestige has remained firm and has expanded ever since; in 1985 Porsche decided to buy all its brake assemblies from Brembo, in 1998 Daimler-Benz followed suit for its S-Class luxury model. At the same time Brembo has seen fit to embrace the newly emerging economies: in July 2005 it set up a joint venture in China to service the main European and Asian auto manufacturers operating production plants in the Far East (www.brembo.com). The LME waxes strong as the world gets smaller.

State-of-the-art LMEs

State-of-the-art medium-sized enterprises came into their own in the 1970s. If LMEs by definition have always been specialized, state-of-the-art LMEs are super-specialized. The immediate cause for this was the emergence of knowledge-intensive industries to service ever more specific demands. At the same time, this new crop of LMEs were in a position to trade off their rarified expertise for a greater degree of independence. In fact, their very establishment was characterized by a relative detachment from obligatory encumbrances because from the outset they had defined, and were subsequently able to control to a considerable extent, their own territories of activity. This does not mean, of course, that the merits of their products were instantaneously self-evident to potential markets, but it does mean that they were able to navigate their course successfully and in short order to the dry land of the big breakthrough. Speed of progress, in fact, is of the essence for these companies, not only regarding their products but in how they build and handle their businesses.

What is more, in many cases they were iconoclasts. Nothing was set in concrete; no practice, be it to do with manufacturing, financing or marketing, was sacrosanct; hierarchies were flattened or turned inside out. They may have been from the outset fabless, for example, not possessing manufacturing facilities of their own. They concentrated on the planning and designing while contracting out the production. In so doing they could dispense with large capital outlays and maintenance expenses – for the days of the garage start-up were rapidly disappearing anyway – and minimize their fixed costs.

Another aspect of state-of-the-art LMEs as a group is that they have arched over the time-span from national-cum-international to global.

When supercomputer maker Cray Research started out in 1972, for instance, its operational vision was probably national plus something. But come the second half of the 1980s, many of them – like Gilat, established in Israel in 1987 – were literally 'born global'. Gilat's product was global: satellites. And its first big break came, not in its home country, but in the United States, the most global of countries arguably. This was in 1992, just five years after its establishment and so demonstrative of the speed at which such companies act, when it got to supply the U.S. pharmacy chain Rite Aid with 3,000 terminals linked to a satellite-base intranet.

Gilat also illustrates another point about the iconoclast. Being 'born global', and so from infancy instinctively aware of the trappings of the latest information and communication technology, including the virtual, allows this breed of company an unprecedented flexibility in how it conducts its global business. Gone is the heavy plod of tentative progression in stages from exports, to licensed distributors, to joint ventures, to 'greenfield' FDI as propounded by the internationalization theorist of yore. Gilat went direct to America off its own bat and now has offices – not factories – worldwide (www.gilat.com). Herein lies another message: while internationalization meant physical displacement, it is less and less an exaggeration to say that globalization can be achieved while *virtually* standing put. In a globalizing world, the supreme expertise of the specialist can render physical proximity to the client a secondary consideration.

Emerging-economy LMEs

To backtrack somewhat in order to fill out the picture, by the mid-1960s two developments were beginning to stir which by the early 1970s would inextricably entwine to give rise to what would become known as the newly industrializing economies (NIEs). The first of these developments was that, with a prolonged period of growth and mounting prosperity among the world's advanced countries, demand for goods was increasing and competition was becoming more acute. This had the effect of inducing Western and Japanese producers to look beyond their own shores for manufacturing chiefly to sustain cost competitiveness. The second was that the first batch of NIEs had emerged to accommodate this trend, having developed in rudimentary form the package required: infrastructure, facilities, enabling agencies, cheap but reliable labor. The Asian economies in this first batch were South Korea, Taiwan, Hong Kong and Singapore, the latter three from early on in this phase in their development displaying an open and liberal attitude towards inward foreign investment not least because of its export-proliferating potential.

Companies formed in these NIEs during this period necessarily tapped into a world industrial economy very different from their German and American precursors a century before. In a mature industry like textiles they could face market constraints imposed by advanced countries. There again, the evolution of an industry and the seeming unassailable superiority of its leading global firms may make joining rather than competing the only viable option. Two aspects have profoundly influenced the character and disposition of firms established and operated under these circumstances. The first is that the Western and Japanese multinationals that dominate most of the key industries which tend to be prominent in these countries have been increasingly looking to manipulating networks to their advantage rather than standing alone. Instead of competing head on across the board, they are formulating collaborative arrangements in specific spheres with other multinationals who otherwise remain rivals. Such strategic malleability compounds risk which is best mitigated by outsourcing. This is where many indigenous firms in the Asian NIEs, in particular, have found the opportunities to grow and prosper. Having proved their worth, they are in a position to participate as subordinate suppliers in such networks which can take the form of what are referred to as international joint ventures or international strategic alliances (Bartels, 2004).

The second aspect, at least as far as Taiwan, South Korea and Singapore are concerned, is the proactive role of government. Being solicitous as to the welfare of their home-grown enterprises interfacing as they were with an increasingly internationalizing environment, they have been active in initiating institutional frameworks aimed at facilitating technological progress, while in the broader sense being instrumental in ensuring the educational foundations for nurturing much-needed expertise and specialist knowledge. Taiwan's Sunplus Technology epitomizes the consequences. This company's inception and subsequent consistent development are both, on the one hand, rooted in a Taiwanese economy set on a course of industrial evolution since the mid-1960s, and on the other, projected outwards to embrace a globalizing market and technological challenge. That is to say, in being founded in 1990 by eight engineers pooling their acquired intellectual resources, Sunplus was heir to a quarter-century of Taiwan's economic development heavily reliant on contract manufacturing and exports, and of politically motivated direction orchestrated by what is referred to as the developmental state.

The Germans and the Americans

In other words, what is being alluded to here is the differentiation between national economies with regard to how LMEs evolve and what could conceivably happen to them based on their particular national origins. Two contrasting images, as suggested already, are afforded by the American and German experiences, which in turn offer signposts for what could be in store for Japanese LMEs. Needless to say, what we are talking about here is tendencies implicit, it would seem, in the business structures of these two countries which encourage companies to react to their environment as they do, while at the same time looking over our shoulder at the rising tide of globalizing pressures. Taking our cue from Dore (2000), we posit the system of Germany as productivist/welfare capitalism and that of the United States as liberal/stock market capitalism. From that we hypothesize that Germany, with its more evenly spread allocation of stakeholder interests, is more likely to seek stability with growth, as needs be, realized by non-hostile merger and acquisition, while the United States, with its insistence on shareholder value, is typified by the necessity for cumulative corporate expansion, of which takeover – hostile or amicable – is an essential ingredient. That is to say, presented in the following are two contrasting alternatives that leading medium-sized enterprises are likely to encounter at accelerating frequency, either as proactive instigator or on the receiving end, as globalization heats up. They are scenarios which Japanese LMEs will be increasingly called upon to contemplate.

Merger and acquisition German style

Behind all this is corporate identity in the eyes of the actor and beholder; how the company came into being, how it evolved historically, what it has become, and what and who it stands for. For many a business in Germany up until the latter part of the 20th century, family, vocation, and geographical location fashioned the mode of operation. The family was the owner, relying on financial input from a bank, and rarely if ever from the anonymous shareholder. The family was also the trade, learned by – not lorded over by – generations of sons in a society where the scientist and engineer were respected as the prime movers, not merely seen as the directors of financial concentration. With such integrated commitment, total or partial divestment of the business could rarely be countenanced. Moreover, the umbilical cord of the small town community was the natural extension of the company's existence, while the proximate surrounds were not unusually host to other companies in the same trade in what amounted to early industrial clusters. Markets had horizons which were only transformed after a long hike over intervening terrain.

However, things have changed a lot since prewar days and the exigencies of survival call for a different bag of tricks. One solution is that employed by KBA as described in Box 3.1, the merging of companies in the same trade and often regionally proximate, not unusually preceded by a courtship of a fair duration. Groz and Beckert provide a relatively early example of this process, bringing together just prior to the Second World War Germany's two largest textile industry needle manufacturers to form Groz-Beckert. In pharmaceuticals, B. Braun and Aesculap traveled along their separate paths for a century or so before B. Braun bought a controlling interest in Aesculap in 1976, a move consolidated by the incorporation of Aesculap into the B. Braun Group in 1998. On the other hand, both Schoen, bringing together shoe machine manufacturers Schoen and Sandt in 1996, and Symrise, the result of a merger between fragrance and flavoring producers Haarmann & Reimer and Dragoco, are the outcomes of latter-day globalization strategies. And globalization, be it noted, has over the past two decades tilted the inclinations of Groz-Beckert and B. Braun towards acquisition both in Germany and abroad, suggesting a convergence of the productivist approach towards the liberal of late. But in saying this, it must be noted, as KBA clearly illustrates, that such global acquisitions have been tightly aligned with established core competences.

Box 3.1 KBA the German

Writ large in one prevailing interpretation of globalization is convergence theory, the idea that business practices will in the process of globalization homogenize. In its history printing press manufacturer KBA, as it has become, can conceivably be perceived to be a particular version of such convergence because of the way it is ultimately addressing the outside world as it internationalizes. But this is nevertheless a German version, which could, like the Japanese version, be a proactive participant in convergence which it is instrumental in shaping as much as being shaped by it.

One thread of this history is the tale of two companies, beholden to and contributing to their local communities, while also being regionally proximate to each other. The other thread, to an extent overlaps with the first while being more modern in context, relating a tale of merger, acquisition and joint ventures. Notwithstanding this recent

---

Box 3.1   KBA the German – *continued*

transformation, however, the heart of KBA's evolution revolves around the two companies of Koenig & Bauer and Planeta. In fact, the relationship of these two was very close from the start in that the forerunner of Planeta was founded in 1861 by a former master craftsman of Koenig & Bauer, not an unusual development in the German business world. Koenig & Bauer itself, having come into existence virtually 50 years before, was just three years after its inception 'taking on the mammoth task of setting up an industrial production line 25 years before the industrial age reached Germany'. So it was, as is by extension KBA, truly a pioneer in the nation's industrialization. This was in Wurzburg, in the German south and to the east of Mannheim, north-west of which is Frankenthal, the original home of Planeta.

The second thread can be traced back to the second decade of the 20th century when, in 1913, Koenig & Bauer acquired Schnellpressen-fabrik L. Kaiser's Sohn of Vienna, this being followed in 1924 by the merger of Planeta with Leipziger Schnesspressenfabrik. Further movement of this sort, though, was delayed until 1967 when Planeta again effected mergers, this time with two other press manufacturers. Koenig & Bauer then went public in 1985 and globalization took hold in earnest with two acquisitions both of American companies in 1990, Koenig & Bauer buying Motter Printing Press Company and Planeta taking over Royal Zenith Corp. The next year Koenig & Bauer increased its stake in Planeta to 75.2 percent and renamed it KBA-Planeta. Come the new millennium, KBA's acquisition strategy was unequivocal, although true to the same pattern of purchasing firms with which it had had close relations. Thus did it acquire in 2001 De La Rue Giori, its partner of 50-years standing, in 2003 Bauer + Kunzi, another alliance partner of many years, and in 2004 Metronic, based in Veitschochheim near Wurzburg, with which it had been collaborating in recent times.

www.kba-print.de

Growing big American style

One of the remarkable things about many companies that can be classified as LMEs is the prolonged period during which many of them have retained that status, in terms of size, scope and core competence, while at the same time at least keeping pace with, and often reinforcing their positions in, their chosen medium. The perennial LMEs and industrialization LMEs cited above are such instances. However, especially within the liberalizing, globalizing environment of the past two or three decades, growth is an almost inevitable corollary of success, as clearly exhibited by KBA. Although, there again, given the added maneuverability afforded by the globalizing arena, this growth may largely be due to overseas accretion rather than domestic build-up.

Association with large corporate size in the interests of survival can, needless to say, be achieved in various ways. The firm in question may simply be acquired by someone else and this is what has befallen some LMEs, evidence that there is nothing intrinsically takeover-proof in the makeup of the LME. The company can be bought by a substantially larger entity as was Binney & Smith, the manufacturer of crayons, by Hallmark Cards in 1984, and likewise Uhde, the maker of chemical plants, by ThyssenKrupp in 1996. They may also be taken over by peers of nearer the same size in a consolidation of effort, as was Loctite by the German adhesives specialist Henkel in 1997, in an exercise bearing resemblance to the mergers effected by KBA.

On the other hand some companies, having passed through an LME phase, have simply become large themselves, although perhaps consciously striving to hold on to a distinctive LME cast. Sony, at least during the phase when it was achieving ascendancy from the late 1950s through the 1980s, is arguably an example of this, with its insistence on not taking the established large firm as a model and its emphasis on small management and an abundant pool of engineers. As for the United States, not a few formerly medium-sized companies have levered their success by way of astute acquisitions, albeit stemming initially from a core competence logic as illustrated by Medtronic featured in Box 3.2. There are two points worth noting here. While this company's domestic build-up would not be unfamiliar to a German or Japanese firm, where Medtronic can be said to be American in its approach is the way it then engaged in merger and acquisition. This is in the interests of both diversification, although incidentally in so doing diluting reliance on the original core competence, and internationalization. Not that acquisitions are unknown to German or Japanese firms on the move – indeed they are being increasingly resorted to, but it is the intensity and extensiveness that American companies practice this strategy that can set them apart.

The second point is the role of the enlisted outsider and what that outsider symbolizes in the company's evolution. Even today the outsider executive appointed to LME-type firms in Germany and Japan is typically required to respect and follow the company's philosophy and traditions. However, when it comes to the United States what the appointment of the outsider in many cases implies is that the company is seeking to gain status within corporate America. For a start, the outsider making his or her entrance is not unusually an external CEO, these days with the express mandate to remold the enterprise's business. But again, this remolding is not so much in his or her own idiosyncratic image but rather based on the ideas and vaunted solutions of a professional corporate mainstream sharing a common basket of formative experiences (Reich, 1992:51–4). Furthermore, more recently it has been observed that institutions other than the firm itself, such as socially prescribed norms of eligibility, investor and media evaluations, and the likes of executive search firms are influencing to an ever greater extent the selection of such executives (Khurana, 2002:186–7). In many ways this creates the antithesis of the particularistic corporate environment a Japanese LME executive expects to work in. However, there may come a time when he will not have the choice. For the ripple effect of the American outsider is worldwide. Although the responses of others have so far been differently nuanced, this has by no means muffled American intervention on its own terms in the globalization game. And the investment fund, the ultimate outsider and the progeny of American business culture is a globalizing tool which, as will be seen, Japanese LMEs need increasingly to bear in mind.

Box 3.2   Medtronic the American

Today Medtronic is a global virtuoso in medical technology. The company achieved this status in three acts as it waxed from small, to medium, to large, central to each act being the pacemaker. When the curtain rises in 1949 the two founders, Earl Bakken and Palmer Hermundslie, have just made $8 for Medtronic's first month of operations entailing the repair of medical equipment in a garage in northeast Minneapolis. By the following year they were representing several equipment manufacturers, but it would take until the mid-1950s before Bakken met Dr. C. Walton Lillehei, a pioneer in open heart surgery from which association came, in 1957, a pacemaker with battery back-up and of portable dimensions cobbled together from parts lifted from other electrical devices at hand in the garage. Upon immediately being proven effective, this rudimentary first step amounted not only to a breakthrough in the treatment of cardiac problems but also the foundation of a core competence the company

Box 3.2   Medtronic the American – *continued*

has nurtured ever since. By 1960 Medtronic pacemakers were in use at the National Institutes of Health in Bethesda, Maryland and other prestigious medical institutes throughout the land and were being exported around the world. The next steps – still as a small company – were, in 1960, obtaining exclusive rights for enabling the production and sale of an implantable pacemaker and, in 1961, the move from garage-apartment set-up to a facility incorporating offices, manufacturing area, prototype lab, library, and auditorium.

It was over the rest of that decade that the second act unfolded: Medtronic became medium-sized – its workforce jumping from 36 in 1962 to 348 in 1968, with sales leaping to in excess of $12 million by the latter year. This was also its LME phase. Having been on the brink of bankruptcy because of the expenses involved in the move as well as for marketing and product development, Medtronic ditched some unprofitable lines and got back to the knitting, thereby restoring a healthy balance and marking up sales of 100 pacemakers a month by 1963. An expanding proportion of its sales of pacemakers and other proprietary devices were to overseas markets, and for this it employed the services of a specialist, Picker International Corporation, to act as sole distributor outside North America. However, in 1968 Medtronic did not renew the contract with Picker, opting instead for vertical integration by gaining direct control of its international operations, complemented by the acquisition of its North American distributors.

The company was laying the groundwork for act three. Other signs were a tie-up with Alcatel of France in the late-1960s, the establishment of specialist neurological and heart valve divisions in the mid-1970s, and the appointment of an outsider as president and CEO in 1976. This president's successor then unequivocally heralded the final act upon his arrival in 1985, with his emphasis on diversification. The strategy encompassed acquisition of other medical technology companies, although this still included pacemaker manufacturers, Vitatron of the Netherlands and TUV of Germany. Moreover, a pacemaker assembly plant was opened in Pudong, China in 1997, and the modestly-priced Champion pacemaker was fashioned for sale in developing countries. Nevertheless, 'by 1999, the company had evolved from what was originally a "one- product" enterprise focused on heart pacing therapy to a large and diverse global leader ...' with some 30 percent of its revenue emanating from outside the United States.

www.medtronic.com

The globalizing LME

So far in this chapter the discussion has dwelt on how non-Japanese firms bearing a resemblance to *chuken kigyo* – leading medium-sized enterprises – have come about and evolved, and how they have changed and coped with change. The reason behind this has been to provide a pool of comparisons for when going over the same ground with Japanese LMEs with the aim of highlighting distinctive national characteristics such as they may be. However, the ultimate goal is to see how Japanese LMEs have internationalized and accommodated globalization, by what means they have managed to impose themselves within arenas external to Japan, and what particular attributes they bring to the table of an arguably increasingly homogenizing planet. To round out this part of the narrative, therefore, we need to consider, with non-Japanese LMEs by way of illustration, the courses firms have open to them when extending their business activities beyond their home territories. One of these methods, merger and acquisition, has already been mentioned on a number of occasions with respect to KBA and Medtronic, who have both resorted to the practice to build up their positions both domestically and internationally. So we shall not dwell upon that again here. Two other main possibilities remain: exporting and going solo.

Exporting

Exporting is the internationalizer's oldest device and it still works handily today. It is important to remember this, especially as it pertains to finished products, because the discourse on globalization tends to wax overly eloquent on foreign direct investment, worldwide sourcing, and intrafirm transactions. In fact, plain exporting has been a beneficiary of globalization too. Vastly improved communications technologies have also worked to the advantage of exporters suitably endowed because reputation and expertise now travel in real time, rendering for some the need for physical implantation of even knockdown facilities in faraway locations probably unnecessary or at least a negotiable option. Furthermore, over recent years there has even been some reverting back to home production and export in preference to FDI for production overseas, especially in advanced industrial countries like Japan, because the domestic attributes such as high-level expertise and mutually beneficial clusters have realigned the odds. More broadly speaking, the smaller, infinitely more accessible world created by the globalizing process has enabled this course of action for both venerable denizens and new kids on the block.

Medicon eG is an example of merging German companies in the same line of business, in this case six of them specializing in the production of surgical tools who came together in 1941 to ensure supplies of metals which were in short supply due to the war effort. This merger inherited an attitude, already nearly a century old at that time and which remains to this day. For when Germany was first industrializing, its products were second-rate, prompting the British parliament to pass the Merchandise Marks Act in 1867, which obliged German manufacturers to label their exports to England as 'made in Germany'. However, the Germans had the last laugh because their products soon surpassed those of Britain, thereby ensuring that 'made in Germany' was a sign of superiority rather than inferiority. Medicon has subsequently been stridently unambiguous, nay stentorian, in trumpeting its strategy based on this claim: 'Every product is made in Germany' (www.medicon.de). And so far globalization has not phased it; Medicon's global presence is still expressed by its complement of some 400 employees purely through exporting.

Going solo

In the literature on the internationalization of the firm it is common to be presented with a progression sequence in which the firm starts with exports, then in order selects distribution or sales agents for given markets, comes to licensing agreements with local manufacturers, and finally invests directly in local production facilities either with a partner or partners in a joint venture arrangement or as sole operator. This scenario can be depicted, particularly for smaller firms, as a somewhat haphazard development reliant as much on chance occurrence as on aggressive go-getting, in which the firm 'satisfices' rather than optimizes because of its knowledge limitations. However, such theorizing is often short on detail with respect to the all-important product. As will become apparent, the LME's activities are concentrated around very specific products which heighten its awareness of market opportunities, both at home and abroad, due to the very fact that the firm itself is embodied within the definition of its core competence output. What this boils down to is that, although the LME's options are far from being boundless, their superiority and uniqueness of product line ensure more flexibility and independence of maneuver. Perhaps virtually everybody wets their toes on trial exporting but, for the LME, the sequence thereafter need not be closely adhered to; stages can be foreshortened or simply dispensed with.

Stihl is a German company which has expanded globally employing its own powerful mix of exports, agents and manufacturing FDI based on proprietary supremacy. In fact, virtually from its inception it needed the outside world to justify its existence. In 1926, Andreas Stihl established a factory in Stuttgart to produce the world's first electric powered chainsaw. By 1971 it could claim to be the world's foremost chainsaw maker. Chief among the reasons for this was the recognition from the word go of the national limitations given its elected core competence. So it went where the big markets were. Already, in 1931, it was developing business in the United States and the then Soviet Union. Operations thereafter expanded over time into Europe, South America and elsewhere. Now, in addition to its seven plants in Germany, it also has factories in the United States, Brazil, China, Australia and Switzerland, complemented by a worldwide network of some 30,000 outlets selling the Stihl line in 140 countries. It has assumed a global persona (www.stihl.co.jp).

Summary

As opposed to large and small firms, little has been written about medium-sized enterprises. Two exceptions are the books published by the American Robert Lawrence Kuhn and the German Hermann Simon. Good reasons for citing them are that they describe firms with similar attributes to Japanese LMEs, adopt an engaging, non-scientific approach, offer a contrast between American liberalism and German productivism, and project a time sequence through the final two decades of the last century (to which this book links up in the 21st). They offer different national contrasts to the Japanese LMEs to be presented here.

LMEs in fact can be found in many countries and have a wide variety of origins. They are presented here under five categories, each illustrated by a particular example. First, is the perennial LME which traces its history back to pre-industrial times and is often still family-run, as represented by winemaker Cordiniu. Second comes the industrialization LME, an individualistic and venturesome outcome of early industrial development spawning diversifying and spreading demand to which companies like toy brick maker Lego responded. Third, is the postwar LME like the producer of braking systems, Brembo, catering to the needs of newfound affluence and growing markets with a taste for luxury. Fourth, state-of-the art LMEs, 'born global' with their super-specializations and iconoclastic solutions are epitomized in the likes of satellite maker Gilat. Finally, there are the emerging-economy LMEs, of necessity merging into the already existing and fast globalizing structure, as illustrated by Sunplus Technology.

German and American modes of LME development are contrasted. Traditionally, the German company has long been associated with family, vocational commitment, and static location. However, in the postwar era, German firms as exemplified by KBA have sought to expand by merger and acquisition, albeit often entailing other companies in the same line of business as well as being, until more recent times, geographically proximate. American firms, Medtronic providing a case in point, have frequently exhibited a taste for growing big. As they have thus expanded they have tended to turn for leadership to outsiders, who bring with them universal norms which supersede the individual corporate culture, the most recent extension of which is the investment fund. Expanding firms these days, whatever their origin, are internationalizing fast. This need not be achieved through merger and acquisition; there is also old-fashioned exporting as insisted on by Medicon eG. Then there is setting up abroad under one's own steam like Stihl.

4
The Stylized LME

Precision and the designated fish

Writing one 100 or so years ago, economist Alfred Marshall opined that Adam Smith and many of the other earlier economists sacrificed precision to the interests of seeming simplicity. They did this by employing a conversational style of writing in an attempt to explain matters which were in fact of daunting complexity. The result was much misunderstanding and time-wasting controversy (Marshall, 1920:30). On the other hand, Marshall was at pains to point out that the social sciences, of which economics is one, are not, due to the nature of their subject matter, equipped in the same way as the physical sciences are to categorize the phenomena with which they deal by ascribing to them a terminology in perpetuity, which is to boot remote from the layman's grasp. Economics must be expressed in terms intelligible to the non-specialist and not rigidly encased in jargon which, in any case, lacks the flexibility to cope with the constant mutations the economic discipline is called upon to explain (Marshall, 1920:43). The commitment to common language when simultaneously seeking precision, therefore, requires a constant attention to ongoing change. In this light, the onus of economic analysis is on the elucidation of differences in degree rather than in kind. It is possible, for example, to envisage firms as being *predominantly* – not wholly – characterized by market or hierarchy orientations.

Our concern in this chapter is what an idealized LME looks like, how it functions and organizes, and what implications this has for its globalizing. This is complementary to the consideration of real-life LMEs from the industrialized world presented in the last chapter before we home in on the Japanese LME subsequently. In his book *What is*

History? E.H. Carr (1961:23) observed that the historian is confronted with an open sea of infinite facts and events. Fishing indiscriminately in such waters serves no purpose. There is no alternative but to decide from the outset on the actual fish deemed desirable for giving definitive shape to the chosen narrative, and then to go for it. Precision, again, is the starting point and here the LME is that particular fish. This amounts to a deliberate and overt exercise in selecting the characteristics of the firm which strike us as being of most interest in accordance with Carr's prescription. It is nevertheless a model because the complexity of detail does not allow us to do otherwise if we are out to formulate a digestible notion (Casson, 1990; Krugman, 1995). 'Economics,' as Keynes (1973:296) averred, 'is a science of thinking in terms of models joined to the art of choosing models which are relevant to the contemporary world.' The aim therefore is to conjoin a set of familiar concepts to impart as precise a definition as possible to the fish we have set out to catch. This can be stated thus: the LME is an *entity* which has undergone an *evolution* while sustaining *motion* and *unity*.

Entity – size, stage and type

Leading medium-sized enterprise is a free translation of the Japanese term *chuken kigyo*. It is interpretive in that it designates a discernible phenomenon. At the same time it is part idealized abstraction and part value judgment, suggesting success in potential and outcome. Actually, the literal translation of *chuken kigyo* would be 'medium-strong enterprise'. However, the way the term is used in Japanese leaves one in no doubt that the firms under discussion have acquired sole or joint leadership in their sphere of business, which in this study is limited to manufacturing. This sphere is often closely defined, notably if it entails intermediary supplies: precision bearings rather than machinery, miniature motors rather than electronics. But the fact that these LMEs are such specialists – and assuming in this case that they are chiefly suppliers of intermediate rather than end-user goods – means that they are possibly servicing both the machinery and electronics industries over a wide range and in considerable depth, as well as other broad industrial categories in the same manner.

By definition the LME is medium-sized, having reached a stage where it is capable of maintaining a relatively high degree of independence and control in its decision-making and execution of business. The LME, therefore, is not a subsidiary nor is it a recent spinoff. It does not have any powerful outsider equity-holder dictating events or being in a position to do so. In addition, the LME is characterized by being intensely concentrated on its core competence, or tightly circumscribed range of core competences, the merits of which it strives to optimize within a recognizably bounded scope of operations. So the LME is an amalgam of size, stage and type, to be elaborated in more detail in the following.

Medium-sized

Firm size is usually assessed in terms of revenues, assets and/or number of employees. Less commonly the size can be related to, for instance, product range, organizational structure, or profit volume. The notion of what is large and what is small can vary with the industry and the national economy. In Japan a firm is reckoned to have become large if it has 300 employees as a manufacturer, 100 as a wholesaler, and 50 as a retailer, or is capitalized at 300 million, 100 million or 50 million yen respectively. Firm size has to be taken into account as part of the background in a general sense because it is one of the key moderators of strategy. The small firm, for example, can be agile and focused but, being relatively strapped for resources, it cannot sustain long-term research projects of extensive dimensions in the way a large conglomerate can (Drucker, 1973). What is more, there appears to be an increasing need to evaluate the right size for the given endeavor. What we are looking to show here is the fact that medium-sized companies, possessed of certain distinctive characteristics, can be successful exploiters ultimately of a globalizing environment. However, at the same time, the nation – the domestic roots of these companies – has far from disappeared and is far from irrelevant to their actions. This is one reason for here fixing the Japanese LME's size at from 200 to around 2,000 employees *within its domestic economy*, although the maximum could well go beyond this depending on the industry and economy as well as the modulations implicit in stage and type.

As for limiting the definition to the number of employees, this has been influenced by the fact that in the case of many Japanese LMEs the figures for employees have been remarkably stable within Japan not unusually for decades while, over the same period, the positioning of these companies within that economy have remained essentially unaltered despite growth in revenues and assets and, for not a few of them, listing on the stock market. That number of employees – from 200 to 2,000 – was the range accommodated by the *chuken kigyo* that Nakamura was reviewing in the late 1980s, and that has not changed much since then either. Taking a worldwide perspective, moreover, where truly large firms have tens or even hundreds of thousands of workers, LMEs thus defined here are still only a small fraction of their size. And stage and type remain to further refine that definition.

As stage/in stage

The stage at which the LME finds itself can be interpreted in two ways. The first is the LME itself as the embodiment of change. In this process it has attained a size and capacity which ensures its independence in decision-making based on a fair degree of organizational and financial autonomy. This is not independence in the absolute sense, needless to say, but the ability to maneuver within an infrastructure of activity, the essentials of which the LME defines and develops for itself. It operates within networks of cooperation, conscious of its ability to bring its influence to bear because of the quality and originality of its products, the abundance of its expertise, and the fact that, as a result, it holds a leading position in its defined area of business which is well beyond the mere filling of a niche with 'something marginally different' (Storey, 1994:11).

Important to recognize also is that, although we are talking here of a stage which in a sense implies a degree of permanence, the term LME is dynamic; it insists on change. Both Sony and Kyocera *passed through* brief LME stages before becoming large corporations. Dynamic behavior is also apparent *within the ambit* of the LME stage; the company can exploit economies of scope and linkage, by extending the applications for instance, to take full advantage of its initially narrowly-based core competence in the early growth phase.

The second aspect of stage is the LME as a distinct phenomenon within a particular phase of industrial development, a stage within a stage as it were, in which the LME is an active agent of change. It was LMEs in the making who, even centuries ago, developed the best writing instruments, medicines and liquors, and they remain to this day having proved themselves adept at both contributing to and adapting to change. There again, not a few LMEs trace their origins to the industrialization phase in their respective societies. Early on they were the blacksmiths who were eventually to turn into specialist machinery manufacturers, for example.

They also straddled the century or so of the battle for supremacy between the artisan and the regimentation of the machine, finally, but not absolutely, won by the latter with the hoisting of the banners of Fordism and Taylorism – the production line and scientific management. Not absolutely that is, because although automation and the rationalized work routine have brought in their wake homogenization, they have also ushered in ongoing diversification because the corollary of the reduced variety implicit in homogenization is the need for greater knowledge and expertise concentrated on a more confined area. What is more, the spirit of the artisan still breaths. And many an LME is to be found at the point where all three – artisanship, homogenization and diversification – intersect in this particular moment in the history of man's organization of economic production.

Leading specialist

So we adopt the assumption that the type of firm designated as an LME is a reality which has derived from complex patterns of social, historical and global evolution supporting its own originality, these being in constant flux through human action and interaction. Within this process the LME has achieved a position of leadership in a tightly defined sector of business, coupled with a marked distinctiveness, which are based both on the character of the enterprise itself and the way it operates, as well as on the products it manufactures and sells.

Key to this pronounced status is the LME's intensive product orientation and coherent and focused strategic precision, presaged on effective innovation through the optimization of product scope and a keen perception of market opportunities. Being the embodiment of achievement and growth rather than the cash cow of short-termism, the LME is at the forefront in promoting employee autonomy and motivation, in that it regards itself as the springboard for creativity and self-fulfillment. This is because it recognizes these employees as crucial to its own definition. That is to say, this definition retains an element of fluidity in that, if the LME is to survive as an entity, modifications to its activities are inevitable over time, and these modifications will emanate in no small way from the reactions to both internal and external events of those who work within and for that LME.

Evolution – background and process

To this day, most firms are initiated by individuals, or small groups of individuals. Not a few of them that have been around for some time have been sustained by the family of the original founder, not infrequently for generations. LMEs, as we have already seen, are quite likely to have started their days and evolved in this fashion. Dramatizing their history a tapestry can be woven depicting the idiosyncratic founder driven by innovative and entrepreneurial zeal, creating new products, applying new processes, employing new materials. In so doing he makes his unique contribution to the elaboration of technologies, industries and markets as he progressively takes on the challenge of a world beyond the physical and psychological framework of the run-of-the-mill small firm. Evolving from this, the family participates, quality is pursued, and core competences are established.

Founder, family ...

At the early stage, the founder is the majority owner, or at least in a position to orchestrate proceedings because the existence of the firm in question is coterminous with his personal capabilities, inventiveness and aspirations. The firm bears the imprint of his character. Around him are arrayed the family members, often likewise deeply involved in the fate of the venture. From among these members will materialize the eventual inheritors and successors to leadership, and indeed the progenitors of further generations of leadership.

However, the term LME is not synonymous with family business. Although it cannot be passed off as merely coincidental that many families have succeeded in not just holding on to but also making good the enterprises established by their forefathers, the LME is not circumscribed by family fiat. It is quite likely that, as the exigencies of expansion take precedence, resort to outside managerial and financial resources may well appear inevitable. On the other hand, evolving as it has the LME has assumed a persona which cannot be simply rejected or ignored. So that, even as family influence declines and the separation of ownership and management starts to take effect, faithful adherence to the company's philosophy and traditions are required of those not of the original inner circle who enter what Gélinier (1996) has called '*la famille élargie*' ('the enlarged family'), or what can be termed the corporate family as opposed to the biological. Firm characteristics are not readily dislodged. Management, for one, remains small and compact, while the organizational structure is still relatively flat with few hierarchical tiers, resisting bureaucratic complexities.

Quality and persistence

In pursuing its métier the abiding leitmotif infusing the LME's every action is quality. Quality permeates the decision-making process; it is demonstrated in product development, manufacturing technology and marketing creativity; it is realized in the superior equipment and personnel; it is the source of the LME's competitive advantage. Aware of its strengths, this kind of firm has a clarity of vision as to its potential, and demonstrates a coherence of approach as it hones its distinctiveness in product, manufacturing process and marketing. It has a demonstrably progressive attitude towards its own development, product innovation, and financial planning. Stated as a formula the LME's *modus operandi* boils down to 'differentiation' *x* 'excellence' *x* 'creativity' emanating from 'originality in products and services' + 'originality in the company's credo and mission' (Kage, 1991:77).

The ideal outcome is that the LME, because of its distinctiveness, attains a considerable degree of independence both *vis-à-vis* its clientele and in terms of geographical spread. Having achieved quasi-monopolies in crafted demand settings it is strong enough to ward off threats from large predators or small opportunists. This in turn is due to its prolonged and exacting focus on relatively stable markets rather than on short-lived trends and transient fashion. At the same time, however, the LME is ever conscious of the environment in which it works and where it stands in that environment, not least what is expected of it and how it is perceived by its customers. In other words, it takes account of its weaknesses as well as its strengths. In sensing its proper level it reins in illusions of boundless horizons and exponential growth.

Core competence

Quality and persistence are the foundations of the LME's core competence. In turn, the LME and its core competence are inseparable. This is best understood by looking at contrasting notions of the firm and what it stands for. Chandler (1990), for example, sees the firm ensuring and prolonging its existence by mastering opportunities which are subsequently deemphasized and discarded as their economic rents decline. Here the *raison d'être* of business endeavor is the survival of the organization *per se* in whatever guise seems warranted at the moment, where the corporate antennae are simultaneously on full alert for the next profitable opening. As against this, whereas, as alluded to already, change is the inevitable concomitant of any enterprise, in hypothesizing on the LME our attention's center of gravity shifts decisively towards how the firm performs with a given accumulation of knowledge and expertise in its core competence at a particular time, as opposed to the depiction of an organization primed to accommodate change and turn it to quick advantage. Neither the LME nor the core competence can be defined without the other; together they comprise a unified conceptual entity. This core competence is embodied in product, process and customer orientation. It is the set of technologies and associated assets that constitute the essential element for the firm's sustainable competitiveness in its chosen line of business. The product, or range of products, is the outcome of the LME's accrued innovative drive.

The possession of effective core competences influences the approach to manufacturing and marketing. The kernel of the LME's proprietary technologies, which embody the distinctiveness of its products, is kept in-house for purposes of both secrecy and cost-effectiveness. Ancillary components and processes are subject to analysis to determine whether they should be outsourced or not. Internally developed technology may well include designated machinery and equipment further underlining differentiation, but also with the intention of meeting specific customer demands. For technology, as reflected in product and process, is inextricably welded to intensive customer orientation as the strategic mode of market entry.

Technology and market are not mutually exclusive, nor is one given priority over the other. Rather they comprise a synergy where quality of output and attention to market needs are of equal moment, rendering the LME's endeavor both technology-driven and customer-driven. Because of the breadth and depth of its product range, moreover, the LME commands a market potential which is not parochially confined; concentrated focus can be combined with extensive operations of worldwide proportions. Hence, the LME's allegiance to core competences can have global implications in a very modern sense, facilitated by the revolutionary advances in communications and transportation.

Motion – articulation

As employed here, articulation is a concept for visualizing the firm's actions and responses in a field of tension between its sustained commitment to a specialized area of business on the one hand, and the constant need to address a changing environment on the other. It focuses on how the firm gains, maintains and elaborates its competitive advantage by aligning and realigning attributes, circumstances and constraints. Competitive advantage is seen as 'a unique position which a firm develops *vis-à-vis* its competitors' (Bamberger, 1994:134), resulting from an accumulation and refinement of activity which at a given point in time demonstrates its superiority to others in a certain sphere. The advantage thus attained is founded on distinctive competences embodied in its unique assemblage of equipment, technology, personnel, information, and marketing capabilities (Porter, 1985). This comprises the proactive toolbox of articulation.

Conversely, the firm is constrained by countervailing external forces, such as competitors and governing agencies, as well as by its own internal limitations, not least its bounded knowledge. The firm that practices articulation, while confident in the viability of its core competence giving rise to its competitive advantage, at the same time demonstrates a heightened awareness of the circumstances in which it finds itself and therefore the need for trading off attributes against constraints to best effect. This is the essence of articulation which is here argued to epitomize the strategic philosophy and evolution of LMEs. To the military theorist articulation represents a strategic philosophy and reality employed both in the deployment and transference of armed forces. It has evolved over time subject to an increasingly sophisticated comprehension of organizational theory as well as group and individual psychology, coupled with advances in weaponry and communications technology. So a military analogy can bring out the key elements of articulation.

Coordination – To commence with the whole unit, the Romans' early contribution to articulation was their disposition in battle, whereby their armies were arrayed in three successive lines each six ranks deep. During the fighting, gaps in the first line were replenished by soldiers in the second line, and ditto in the second line by the third. This articulation allowed for a maneuverability and responsiveness lacking in the traditional phalanx. Here articulation entails the intentionally contrived and closely combined movement of a self-consciously coordinated group.

Focused leadership – Still with the Romans, the phalanx permitted the general, from his vantage point on an eminence, to command the whole army rather than only part of it. The group is therefore piloted by a conductor orchestrating activity while benefiting from the elevation of a global perspective.

Solidarity – By contrast, the Swiss militia unit of the 14th century comprised manpower from the same village, town or guild. United by kinship, neighborhood and occupational ties, the men fought on foot with no body armor or shields, their sole weapon being a halberd, which is essentially a combined pike and axe. As opposed to the conventional battle line, they adopted a formation resembling solid squares. This allowed the men to stand shoulder-to-shoulder and level their pikes in all directions to defend flank, rear, and front. The roughhewn solidarity propelled their success.

Decentralized initiative – By the end of the 18th century the revolutionary French were elaborating on this rudimentary prototype as they experimented with a much larger armed force of separate, self-contained, and interchangeable parts. Although subdivided into divisions, these separate units still remained component parts of the whole army, acting together and capable of quickly reuniting. The subunits were decentralized, while at the same time the army as a whole was more compact and cohesive. Thus do we see a manifestation of articulation of the whole organizational entity as represented here by the army, plus articulation *within* that entity as demonstrated by its divisions. Independent initiative was subsequently devolved down to the battalion which, as part of the articulated division, became a basic unit for maneuver. Such devolution ensured better cooperation among individual soldiers and greater motivation, conscious as they were of the support of their comrades and loyal as they were to the group and its standards.

Technological progress – The introduction of the portable battery-powered radio in the first half of the 20th century had two complementary and intensifying effects. By allowing for instantaneous contact it reasserted and strengthened central control, while simultaneously it increased the potential for articulation on an unprecedented scale and depth as it was diffused down to tank and squad. The unified entity was increasingly refined, while overall coordination, central command and solidarity were enhanced in the process (Jones, 1987).

Unity – the chamber

Thus far we have discussed how articulation functions; now we turn to what holds it together. The new metaphor is the 'chamber'. The notion of the chamber is employed to symbolize the essence of the LME in terms of structure, aims and strategy with the intention of capturing the LME in time and space. That is to say, the LME as described in these pages is a particular phenomenon of business organization prevalent notably in certain advanced industrial countries in the current era. It is still embedded in an institutional framework over which it has more or less influence, but its individual structural reality is undeniably tangible and visible.

Lean structure

The chamber is lean in executive temperament and sparse in bureaucratic hierarchy; it offers an arena throughout for individual responsibility, personal satisfaction and commitment. Functioning as an entity it is the proactive body which interfaces with the exterior. Within it are creative and reactive mechanisms seeking to find innovative solutions within the bounded scope the chamber affords. Undergirding these mechanisms is proprietary knowledge accrued in concentrated depth and breadth which guarantees distinctive competences and raises the drawbridge against rival encroachment.

Distinctive objectives

Structured thus, the stylized LME as chamber aims for self-defined distinctiveness in product and service, optimization of internal operations, external interfacing and a high-tension balance of offensive and defensive ploys. With respect to the first of these, the LME looks to forge its own unique image through original product or service innovations, the implications of which it is in a better position to understand than anybody else. This is, needless to say, not a complete understanding, but a bridgehead from which it is better placed to view the possibilities. At any event, this allows for detecting potential customers before those customers even realize they are in the frame. Hence did Mabuchi divine new users for its electronic motors, for example.

As for optimization, the LME is not intent on venturing to the margins of internalization. Rather, its optimization energies are directed much more emphatically to extracting the most from what it knows best. The chamber then provides the protective time and space within which plans for offense and defense can be laid. For concentrating too much on defense would frustrate creation, while an overemphasis on attack enlarges the risks. To succeed, both attack and defense must be positively addressed to achieve the necessary balance, which is of high intensity because it is presaged on simultaneously attacking hard and defending hard.

'Emerging strategy'

The chamber then evokes an interior which extends scope for innovative and sustained industry animated by assertively mobile and flexible mechanisms, while encased by an outer circumscribing shell. Looked at as a whole, the broad outline of the LME's actuality and activity is tolerably discernible. However, the precise details overall are not clearly enunciated because they are in constant flux within the interior. Strategic choices are then acted upon as latent possibilities are identified and addressed. This is what Mintzberg and Walters (1985) have called 'emerging strategy' which evolves to ultimately realize consistent patterns as a result of the coalescing activities of the participating individuals. The firm's competences and competitive advantages are mapped against circumstances and the resultant perceived opportunities as they arise. Its area of expertise being concentrated as it is, the LME applies its ingrained technological capabilities to interpret demand in its own fashion, aimed at tightly delineated target markets in an exercise Porter (1985) has referred to as 'focused differentiation'.

Figure 4.1   The Stylized LME

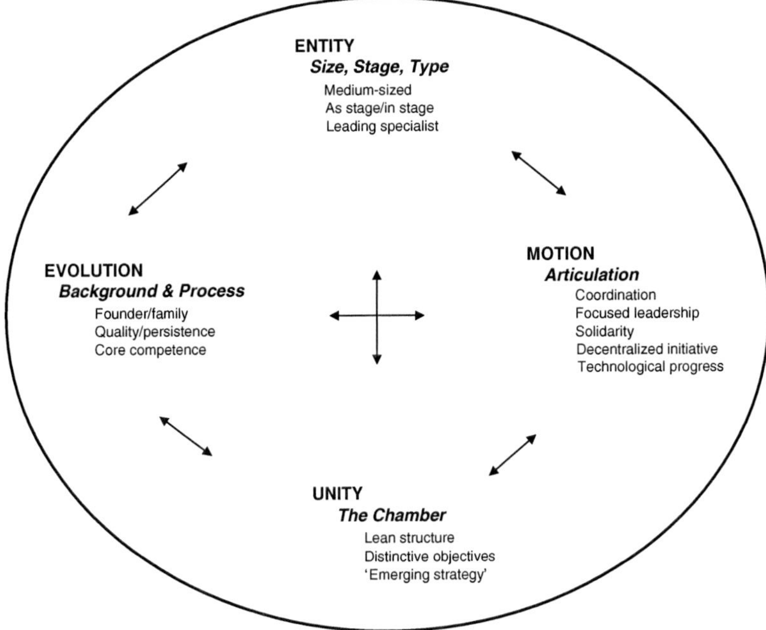

The stylized LME and globalization

Different fish have their singular preferences and capabilities. Likewise, firm-specific characteristics decree behavioral options. It follows, therefore, that different types of firm are capable of developing their own distinctive perspective on globalization. Not being a behemoth the LME cannot carpet the planet with the magisterial presence of a great multinational. Given its bounded circumstance in size and scope, the temptation could be to place its strategic options somewhere midway between the small firm and the large corporation. Except for one thing: the modern LME with its superior core competence has locked into a natural ally in globalization. For a start, with its proven track record, erected entry barriers, and unique firm-specific characteristics it stands out from the crowd. It can be seen. With its breadth and depth of capability and heightened awareness of its own potential it can move in on its customer base on its own initiative. With its built-in competitive advantage it can cultivate latent opportunities in industrial and commercial clusters and regions worldwide.

Complementing this is the exponential speed of globalization, especially over the past two decades, enabling the real-time acquisition of information, transmission of designs, and conclusion of transactions. The sinews of global movement are increasingly pronounced. What this means is that the LME, with all its advantages, can articulate a strategy which pinpoints with far more accuracy than 30 years ago where it wants to invest in manufacturing, who it wants to serve and how, what routes it will adopt to best accomplish its profit-making mission. Although this bears in mind, of course, that this is against a backdrop of intensifying competition where rival solutions threatening to undermine their existence are incessantly in the wings. Life is not easier or simpler, just more up-front. But the stylized LME is in its element.

Summary

The LME is a particular category of firm, the defining elements of which are 'medium-sized' and 'leading'. Thus defined it is abstracted from other firm categories for analysis, which entails viewing it in terms of size, stage and type (entity); background and process (evolution); articulation (motion); and the 'chamber' (unity). Firm size can be seen in terms which include revenue, assets or number of employees, for example. Here the latter is employed because it has been the most consistent over recent times, especially in Japan. Stage is related to change which is intrinsic to the LME. First, the LME is the embodiment of change in that it is its own decision-maker, dynamically responding to external forces and being dynamic within itself, in a state of sustained entrepreneurship. Second, it is the agent of change within a particular stage of industrial development. It is the constant specialist, differentiating even as homogenization progresses. As a type of firm, it leads a business sector based on a marked distinctiveness, and values its employees as agents of essential change.

With respect to background and process, not being synonymous with founder and family, the LME can undergo separation of ownership and management and yet remain compact with a minimum of hierarchy. Quality and persistence are the foundations of its core competence and the LME and core competence comprise a unified conceptual entity. The way it functions can be characterized as 'articulation', or the trading of attributes against constraints. The essentials of articulation are summarized through a military analogy: coordinated strategy, focused leadership, solidarity, decentralized initiative, and engagement with technological progress. The 'chamber' then epitomizes the LME as a tangible and visible unity. It is the proactive body interfacing with the exterior while also defining the potential scope of activity.

Given these heightened aspects of function and form, the globalizing LME has a more focused approach based on its core competence, its intricate knowledge of its specific business environment, and its established reputation. Globalization is its natural ally because the speed of communications and information transfer enable the LME to articulate an even more accurate strategy.

5
Early Industrialization and the Adumbrated LME

LME stirrings

We have established in Chapter 3 the universal character of the LME. However, having said that LMEs are not unique to any particular national economy, there is still leeway to entertain the proposition that certain national economies can be distinctly amenable to the germination and spread of this kind of enterprise. Japan, it is argued here, is one such country. And furthermore, it is contended that this eventuality has been instrumental in bringing a distinctive quality to how the Japanese economy has addressed the outside world as it has internationalized and globalized. What Japan brings to the table of globalization has its own unique presentation and flavor based on the inimitable particulars of its preparation. The objective of this chapter, therefore, in the first instance is to probe the background and, in so doing, to unveil the scope opened up for the evolution of the Japanese LME by describing the surrounding historical circumstances.

Such a briefing can conveniently commence with Japan's first encounter with the irresistible force of aggressive Western demands for access in the 1850s, the rude awakening which set Japan on its course of modern industrialization. This is the Japan oft cast as a 'late developer'. But having for some centuries already honed their own arts of administration and craftsmanship, the Japanese hardly confronted the challenge empty-handed; the samurai and the artisan, the administrator and the entrepreneur, were both in a position to react in their respective fashions. Response, in other words, found its expression at various levels of Japanese society and was, moreover, in short order disseminated broadly over the archipelago. The fact that this response was far from being simply top-down and that artisanal talent and inclination were widely dispersed throughout the country would have implications for the launching of putative LMEs right from the start of the Meiji period through to the Second World War and then beyond.

The samurai and the artisan

Pre-Meiji Japan did comprise an elaborate social hierarchy, however, and it is with a view to attempting to understand the responses and actions at the various levels that we employ the concepts of the samurai and the artisan as representative of the two polar reference points. This is while bearing in mind, needless to say, that contemplating the two poles is with the objective of better appreciating the gradation of the spectrum. While they were, moreover, very much part of the scene during the Tokugawa period which preceded Meiji, the samurai and the artisan are also employed as abstract concepts to help elucidate subsequent economic developments in Japan.

The samurai administrating

By the time the American Commodore Perry appeared with his four black steamships in Edo (Tokyo) Bay in 1853, the Tokugawa Shogunate had ruled the country for well nigh 250 years. One of the key elements enabling the prolongation of the regime's power was the role of the samurai class. Having been the warriors who battled for their lords through to the beginning of the 17th century, under the victorious Tokugawa regime the samurai were converted to civil administrators. Comprising from five to six percent of the nation's population, they were eminently suited to function as such, being both few enough to constitute an easily distinguishable elite, and numerous enough to infiltrate an authoritative presence from the capital of Edo out to the remotest domain.

The samurai were destined over a span of two centuries or so to create a viable and sustainable civil administration. As bureaucrats they recognized the need for education so that, as the level of education increased, government became more impersonal and legally defined, and the qualified acceptance of meritocracy as a standard slowly emerged. However, although scholarship and ability were recognized as essential for good government, status – both of the samurai as against the non-samurai and within the samurai class itself – remained paramount. This status amounted to a collective identity based on an honor culture which placed the samurai apart to the extent that only they were deemed to possess a sufficient sense of responsibility to govern (Ikegami, 1995). But even as the samurai class consolidated its reconstituted position it sowed the seeds of its own transformation through an institution having its attractions for both the warrior and the bureaucrat, although difficult for either to control. This was Confucianism – or neo-Confucianism to be more precise – which the bureaucratic Tokugawa samurai adopted as its guiding rationalism for peacetime administration. In its Japanese guise, Confucianism emphasized a warrior-type loyalty to one's superiors; more generally, it infused the principles of meritocracy into bureaucratic proceedings. Loyalty supported the props of status, while meritocracy leveraged substitution and change. Moreover, the very fact that society's upper stratum laid stress on this body of ethics and teaching attracted the lower orders to whom it became increasingly accessible. The prosperous and intelligent artisan could follow the same paths of instruction and perhaps enter the realm of the elect, as was more and more the case as the Tokugawa era wore on. The result was that, while the physical pedigree of the samurai class may have been progressively diluted, its precepts and beliefs, as well as the perceived need for its services, were left intact. The samurai remained.

The artisan at his station

During the Tokugawa period, then, the samurai class was at the pinnacle of a society of multiple inherited gradations. Inherent in this was a vertical stratification from peasant to village headman to *han* (domain) bureaucracy to daimyo, the latter being the head of a domain which had typically – although by no means inevitably – been passed down through generations. So the humble artisan could theoretically trace a hookup to the top of the pile. The outsider in this artisan's set of references was more horizontally positioned: a person from another village or engaged in another trade.

Tracing a conventional outline of the artisan's position can go as follows. For the individual artisan in Tokugawa Japan and the group to which he belonged this vertical configuration came with constraints in the form of oppressive social norms, such as restrictions on land sales, and specific obligations according to occupation and status. However, on the other hand, occupation and status were positioned in designated frameworks subject to degrees of autonomy as deemed appropriate, although always deferential to the next higher rung of authority and attentive to a complex matrix of obligations. This was coupled with communal responsibilities, that is, consciously active participation in furthering social welfare and improvement. Thus did villages collectively maintain their own utilities, enforce by-laws, police their territories, and impose local taxes, for example (Whitley, 1992). The village was allowed such leeway, moreover, because, like the trade guilds and the merchant associations in the towns and cities, it was part of a compact conjunction of what were essentially vertically integrated social components. Control ultimately resided with the shogun at the apex. The high level of vertical integration then having been achieved permitted a high level of decentralization as illustrated by mediated village autonomy.

The scope of autonomy granted at the local level, on the other hand, was in turn instrumental in allowing for the development of non-agricultural rural industries from which individual households could profit. What is more, the decentralization of such activity occurred not only in a national context but within the domains themselves. Craftsmen had been restricted to the domain capitals at the beginning of the Tokugawa period, but by its end the making of silk cloth, for example, had spread out to the rural farmhouses with machinery and materials supplied by local wholesalers. Instead of being concentrated in a few major cities artisanal expertise – *monozukuri* – was dotted all over the country (Morris-Suzuki, 1994). Furthermore, conjoining these two contrasting pictures of the artisan's existence leads us closer to what he represented. Conformism there certainly was. But within that there was still scope for an individualism expressed in autonomously motivated action and self-reliance, the very essence of grass-roots industry as it was to develop in Japan. One consequence of this is the way firms which were eventually to become leading medium-sized enterprises, as frameworks within a graduated structure, were left the space to grow and prosper. Independent to the extent their specialization conferred, they still owed a fealty to the unity. Denizens in the main of the decentralized territory of the artisan, they were nevertheless implanted in a structure invigilated by the samurai.

'Late developer' Japan

Japan was neither feudal nor particularly backward in its governmental and commercial organization by the end of the 1860s; what it lacked was access to western science and technology. The vertical linkages, as symbolized by the constant comings and goings of the provincial lords from Edo (Tokyo) to their domain capitals and back again, had carved out patterns of communications nationwide which generated market activities along the routes and increasingly specialized production in those provincial capitals, or castle towns as they were called. Moreover, at this local level the administrators – principally samurai – busied themselves with the stimulation of domain agricultural production, the regulation of regional markets, and the movement of goods within and beyond their territories. During the Tokugawa period, a national economy began to take shape predicated on a progressive convergence of language, thought and practice, along with an elaboration of regional and proto-industrial specialization.

Nevertheless it was still 'late' in the eyes of the 'advanced' Westerners now banging on its door. 'Late developer' can be seen as a term which condemns, describes, and orientates. Condemnation is presaged on external judgment; description is a neutral assessment; orientation is the internally-generated response. Condemnation as judgment derives from power based on superior technologies and organization and the assumption stemming from this that the one possessing the greater power is justified in dictating the institutional framework the less powerful should adopt, which is needless to say, a facsimile of its own. In so doing it straitjackets theory of economic and industrial development, by implication dismissing alternative routes.

The observer of mid-1860s viewing Japan *vis-à-vis* the Western powers would have noted that it had no steamships, telegraph, concentrated manufacturing with factory-disciplined labor, nascent mass markets, modern military, or merchant marine supported by internationally operating insurance and banking services. The bulk of the population would be still on the farm offering little or no premonition of mass migration to industrializing urban centers. Infrastructure in terms of transportation, communications, public utilities and order would be rudimentary. Productivity would be low and organization relatively primitive. The West, on the other hand, in indicating what this 'late developer' lacked provided the orientation reference.

Orientation here means exercising the art of the possible having taken stock of the options: bringing in the new technologies, adapting native techniques and skills, leapfrogging stages which no longer need to be passed through, and protecting the early outcomes of incipient industries from extinction by stronger, more experienced rivals. Rigorously adhered to this is a forced march, mobilized by different attitudes and perceptions from those who have gone before. Institutions and methods are more channeled into grooves directed at visible ends than spontaneously adapted to emerging definitions. Just for the very reason that the 'late developer' locks in part-way, laden as it is with its own cultural and economic accoutrements, the paradoxical likelihood is that its course of development will differ, perhaps quite significantly, from the pioneer even as it attempts to emulate (Gershenkron, 1966).

The samurai's consuming preoccupation in all this was to concentrate on what was missing, or to what was needed to catch up, these being largely embodied in what has to be brought in from outside. Concealed, or more accurately perhaps, receiving less attention as drivers for progress, were the native capabilities manifested through the artisan. Behind the smooth lines of infrastructure building – of the new railways, shipping lines and steel mills – was a world scrambling in its turn to redefine itself against a fast-transforming backdrop, and in so doing learning new skills and techniques and adapting them to what was already there. The mind of the artisan was no doubt far less preoccupied with the broad sweep of his country's late development than with, say, applying traditional techniques to the production of industrial ceramics, but that in no way diminishes his significance in the overall outcome. Moreover, the way he interacted with events influenced the shaping of the nation's industrial structure in its own distinctive fashion.

As the Meiji Restoration and the concomitant industrial revolution took hold in the last quarter of the 19th century, therefore, the newly constituted bureaucratic samurai intensified their penetration of the provinces aided by the vertical linkages rooted in the country's Tokugawa heritage. But the graduated, corresponding frameworks also still functioned in their turn. In the dissemination of information concerning new technologies and processes, for example, the central ministries coordinated and partially funded operations carried out by the prefectures. Within the latter were inaugurated local research laboratories and technical high schools intermediate between research institutions in Tokyo on the one hand and the village or region on the other. At his local level trade associations and the like provided the loci in which the established traditional artisan and a putative new breed of businessman – frequently a hybrid of both – could exchange ideas and absorb the information emanating from the center.

The logic of Tokugawa-style social control had radiated native *monozukuri* in the form of crafts, construction, food processing and brewing first out to the castle towns and thence to the rural communities, and these now presented the reception points for *monozukuri* from abroad. 'The effect was a system in which new and imported ideas were not simply concentrated in a few elite institutions run by the state or large private enterprises, but rather dispersed throughout a wide range of bodies, varying in size, structure and geographical location, but all involved to some extent in the processes of importing, modifying or developing new technologies' (Morris-Suzuki, 1994:104). So from the very inception of Japan's modern industrialization a large number of people were involved, many armed from the start with at least the rudiments of education along with experience in a trade and therefore positioned to react positively and with alacrity to the new economic opportunities created by Meiji government reforms. This was a symbiosis. The samurai could not have planned and implemented on the national scale as he did if the artisan had not been ready to respond positively and more or less in symmetry.

In this coordinated effort, distance, adaptation and proliferation were key factors. Despite the perceived economic and political threat from the West, Japan was still sufficiently geographically removed from Great Power interference to develop a large range of locally-made, 'modern' goods for domestic consumption in relative isolation, without the need to compete head-on with foreign-made goods. This offered considerable scope for what came to be referred to as 'Meiji technology', the practice of combining the best traditional techniques with imported technologies incorporating modern science. This in turn was facilitated by the fact that even the technologies imported from abroad, compared with nowadays, had a trial-and-error element about them while requiring relatively lower levels of capital, education and skill (Doane, 1998). Not being that far behind, the Japanese could imitate, modify and transform over a broad spectrum, if not in the state-of-the-art spheres of advanced and large-scale machinery and equipment. Thus the last quarter of the 19th century saw a veritable proliferation of entrepreneurs setting up factories to produce simple, light industrial goods with a ready urban demand such as ceramics, paper, food and beverages, intermediate goods represented by textiles, cement and agricultural fertilizers, and extending to time pieces, elementary machinery and electrical motors.

Meiji entrepreneurship and the individual entrepreneur

To this day, firms founded during the Meiji period still remain intact, with their origins easily recognizable. Some of them can now be designated as LMEs and a small selection is presented below. But before that, another key factor must be introduced: the individual, and in particular, the individual entrepreneur. In casting doubt on the viability of certain explanations of Japan's industrial success in the 1970s and 1980s, business guru Ken'ichi Ohmae opines: 'it is not Japan that was so stunningly competitive, but only a handful of industries within Japan and, more to the point, only a handful of companies led by strong individuals within those industries' (Ohmae, 1995:65). The stress on *individuals* is taken; it is a welcome antidote to the group-cum-bureaucratic depiction of Japanese organization all too often proffered.

During the era under discussion, for example, the forerunner of Toshiba was set up by inventor and entrepreneur Tanaka Hisashige in 1875 to manufacture telegraphic equipment. Hattori Seiko, later simply Seiko, first saw the light of day on the initiative of Hattori Kentaro and two partners to make wall clocks by amalgamating Tokugawa metal casting and smithy techniques with the latest watchmaking methods and designs from Switzerland (Fruin, 1992: 71, 103). Tanaka was essentially a government contractor contributing to the build-up of the nation's import substitution to avoid over-reliance on foreign equipment. Hattori was an independent urban entrepreneur, epitomizing the conjunction of home-grown knowhow and sophisticated foreign inputs. But in both cases it was their stellar individual attributes which galvanized returns.

Where one can take issue with Ohmae, however, both with respect to the contemporary Japanese economic scene and the dawn of its industrialization, is his parsimonious assessment of the country's mobilizing forces. It is in essence an elitist samurai argument. Admittedly, it no doubt included 'strong individuals' of artisanal provenance like Messrs Honda and Matsushita, both still active in the 1970s in two of the 'handful of industries', automobiles and electronics. However, it falls well short of appreciating the depth and variety of increments contributed by self-motivated individuals and the firms they have created and run, and, for that matter, the makeup of Japan's industrial structure as it has evolved over the past one and a half centuries. To quite an extent, the large concerns Ohmae evidently has in mind are strong only because the underlying and surrounding foundations are strong. These foundations are in the main the province of smaller firms, often the suppliers to larger ones although not unusually partially or completely devoted to end-user products. In an economy where as little as 20 percent of the automobile is made in-house by the brand-name assembler, the headlights, seats, and electronic devices, just to take from the top, come in from outside, in turn contrived from parts and materials outsourced by these component suppliers.

Moreover, in an economy which adjusted rapidly to the exigencies of change and development, retaining much of its own while assimilating from abroad the ideas and physical conveniences traveling in the wake of modernization and mounting affluence, the populace was able to buy appreciably more *sake*, processed seaweed and ceramic ware, as well as adding fountain pens, bicycles and three-piece suits. The Meiji period witnessed the unfurling: traditional industries extended and adapted; new industries jumped on the bandwagon. Individuals and their own agendas were indispensable ingredients in Japan's industrial evolution. 'The modernization of Japan was not simply a matter of top-level changes. It was also a cumulation of a mass of small initiatives by large numbers of people who could appreciate new possibilities, make new choices, or at the very least allow themselves to be persuaded to do for the first time something they had never done before' (Dore, 1965:104). Or in our terms, the samurai's ambitions for his country would have been seriously compromised had they not benefited from the support of the artisan's complementary aspirations.

LMEs from Meiji

Firms which are now classifiable as LMEs counted among those incorporated during the Meiji period, some for the first time as responses to industrializing demands, others with historical roots wending way back into Tokugawa feudalism. Some, although not all, owe their birth and early gestation to individuals, some of which again remain to this day welded to founding family interests. In the following we introduce the initial set of examples, these being firms which were incorporated during Meiji. This will be followed by further sets of LMEs originating in sequential historical stages to current times. The sequence is 'Japan-oriented', so that it is not identical with the groupings used for the universal LMEs introduced in Chapter 3. Nevertheless, there are many similarities which will be alluded to.

For a start, with reference to the universal LME designations, it may be noted that of the six companies established during the Meiji period in the above list, Koransha, because of the continuity in character of its development from 1689, would be classified as a perennial LME. The rest would go under the heading of industrialization LMEs, while taking note that Danto's initiating tableware and subsequent tiles owed a debt to the lustrous glaze of the vases traditionally crafted on the island of Awaji in Osaka Bay, and Nippon Paint had been preceded in Japan by centuries of experimentation in and perfection of lacquer and pigments. Both had now embarked on a modernizing course, Danto switching to tiles at the turn of the century, intent on achieving the standard set by the West and in so doing substituting for imports, Nippon Paint taking the national lead which it claims to hold to this day in the development of the manufacture of Western-style paints.

Figure 5.1 LMEs from Meiji

Company	Established (Founded)	Location	Product Range Initial → Current		Stock Market	Founder Family
Shinagawa Refractories	1875	Tokyo	Firebricks →	fireproofing	Section 1	No
Koransha	1879 (1689)	Saya Pref.	Ceramics →	quartz glass	Unlisted	Yes
Nippon Paint	1881	Osaka	Paint →	chemicals	Section 1	No
Danto	1885	Osaka	Tableware →	tiles	Section 1	Yes
Tokyo Rope	1887	Tokyo	Manila rope →	steel cord	Section 1	No
Owari Precise Products	1906	Nagoya	Clocks →	precision screws	Section 1	Yes

NB: 1. Established (Founded): It is not uncommon for Japanese firms to distinguish between when they were set up by the original founder and when they were incorporated in their modern form.

2. Product Range: This is intended to give a rough idea of the scope of the firm's evolution since it started out.

3. Stock Market: The Tokyo Stock Exchange is the dominant stock market in Japan, followed by that of Osaka. The Nagoya exchange is as a distant third. All three are divided into two sections depending on capitalization and other qualifying factors.

4. Founder Family: This indicates whether members of the founder family are currently participating in the running of the company as top shareholders or senior executives.

Shinagawa Refractories and Tokyo Rope are quintessential instances of response to the exigencies of an industrializing Japan, the former at the outset as the first private-sector firebrick maker in the country. The latter, being the first in the field with industrial Manila rope in 1887, was likewise the domestic pioneer with wire rope a decade later.

Individuals and families were prominent in some of these ventures, evincing both the heroic romance in the challenge of the moment and the resort to practical solutions. Enthused by the notion that the gaslight symbolized the civilization and enlightenment (*bunmei kaika*) Japan was then busy assimilating, Nishimura Katsuzo, the founder of Shinagawa Refractories, devoted himself to his dream of realizing the domestic production of firebricks to be incorporated into gas generators. More prosaically but no less determined, the Moteki brothers, Haruta and Jujiro, combined with Nakagawa Heikichi, chief engineer

of the painting team in the Imperial Navy dockyard, to form
Komyosha, the forerunner of Nippon Paint. Since then the company
which became Nippon Paint in 1898 has followed a straightforward
trajectory of specialization, accumulating through intensive research and
development a boundless store of knowledge and information about its
core competence of paint products and coating systems for anything
from automobiles to construction to electrical equipment, veering but
slightly into related fields of chemicals and materials. This is the kind of
logical extension that Shinagawa Refractories pursued also. The aftermath
of the Russo-Japanese War in 1905 saw a boom in construction, encouraging it to add decorative bricks to its portfolio, while still remaining faithful to its original calling of fireproofing. The result was that its
distinctive red facing tiles found themselves gracing the exterior of the
original central Tokyo Station building first erected in 1914.

The tale of Owari Precise Products, on the other hand, describes a
transformation in competences. Actually the company started out in
1906 as Owari Clock because that is what it made, as such being one of
those early contributors to much-needed exports. Just over 30 years
later in 1938, however, it was establishing a new plant to manufacture
screws for military aircraft, thus becoming one of those Japanese firms
whose singular expertise was appropriated to address the inevitability
of looming hostilities, which in the process irrevocably changed its own
course. Thenceforth, during the postwar era in the 1950s and 1960s,
Owari Precise Products, as it was to become, turned its forging skills to
supplying the automobile industry with screws, transmission rings and
other parts on a path which was ultimately to lead to the inauguration in
2002 of facilities for manufacturing screws in Columbus, Indiana and
synchronizer rings in Bangalore, India as its globalizing effort took off
with the new millennium.

Box 5.1   Koransha the venerable

Turn to the company outline in Koransha's website and you will be
at once taken by the elegant Western-style cups and saucers in
turquoise blue with orchid motif. Go on to the company history
and you will immediately be assailed by a bowl and its lid in the
subdued profusion of orange, gold, white and sky-blue of vintage
Arita-yaki. But cast your eyes down to the bottom of the outline
page and you will see that, as well as plants for producing *objets
d'art*, Koransha also runs a factory for making insulators. Return
once more to the history and you will note that, while steeped in

---

Box 5.1   Koransha the venerable – *continued*

the tradition of its Tokugawa origins and wedded to customs inherent in Japan's cultural heritage, from the very start of the Meiji period Koransha has been in the thick of the country's industrialization endeavor. From the very outset of the modern era, in fact, Koransha has traveled on a parallel course of old and new, bridged by its expertise in ceramics. It is a perennial LME for which a venerable past still holds the key to the future. To boot, it presents a singular epitome of Japanese experience and practice.

This epitome can be related under the rubrics of sustained tradition, modernization in response to a national agenda, and extending international reach, all three of which are interlinked. To commence with tradition, Koransha was founded in 1689 by the first Fukagawa Eizaemon to make ceramic ware, 73 years after – so the story goes – a Korean potter named Risanpei discovered white china at Izumiyama in Arita which he used to become the initiator in firing porcelain in Japan. Thereafter Arita-yaki was to enjoy the patronage of its domain in Saga, Kyushu until this privilege was forfeited to modernization following the Meiji Restoration in 1868; while the inherited name of Fukagawa Eizaemon persisted through ten generations to the Second World War, and even now a Fukagawa is president – Shin'ichi at present. Symbolic of Koransha's persistent ties to tradition throughout the modern era, moreover, is its appointment as of 1896 as manufacturer of tableware by the Imperial Kitchen, a bureau of the Imperial Household Agency. In addition, the company saw fit to reassert its continued commitment to its original core competence with the opening of a new ceramic arts factory in Saga on the occasion of its 300th anniversary in 1989.

But that was by then virtually 120 years after Koransha had started on its modern ways. For as early as 1870, Fukagawa Eizaemon VIII had received word from the samurai, in the form of the Ministry of Industry's telegraphic communications bureau, to attempt to manufacture insulators, in which it succeeded in short order. This association with government agencies has persisted throughout: in 1908 it was building kilns as designated by the Ministry of Agriculture and Commerce; in 1937 it was benefiting from research subsidies from the Ministry of Commerce and Industry; in 1999 it was consigned by the Ministry of International Trade and Industry the task of putting together an R&D program based on a regional consortium. Running parallel has been its evolving dealings with the

Box 5.1   Koransha the venerable – *continued*

private sector coupled with progressive product development on its own account: by 1913 it was inaugurating deliveries of high-pressure insulators to the likes of Tokyo Shibaura, now Toshiba; come 1970 it had completed a special ceramics factory within the insulator manufacturing plant, looking to the accelerating expansion of the electronics industry, and was starting to come out with its steatite ceramics.

Furthermore, Koransha was anything but a slacker on the international scene. In fact, in the first instance, its traditional Japaneseness was a positive boost, leading to a certificate of merit for its porcelain at the Centennial International Exhibition in Philadelphia in 1876, and over the subsequent 50-odd years to similar awards in Paris, Seattle, London, San Francisco and Liège. Then, nearly a century after that initial success with the old, having completely revamped its facilities for the new, in 1962 Koransha was launching its insulator export drive in concert with the commencement of the high-growth era which was to establish Japan as a pre-eminent industrial nation.

www.koransha.co.jp

Summary

Japan is distinctly accessible for the generation of LMEs and some of the reasons for this can be derived from its historical background as seen here from the polar standpoints of the 'samurai' and the 'artisan'. In economic terms Japan has been characterized as a 'late developer', a designation which can be viewed in three ways. First, it is a condemnation based on external judgment. This was the reference for the samurai, who as a consequence downplayed, while not dismissing, the indigenous achievements of the artisan. Second, it is an objective description of what is lacking. Third, it is an orientation for taking stock of the options. For Japan this latter translated into a planned campaign of initiating, guiding and protecting, orchestrated by and favoring the samurai but with a marked artisanal input.

This hierarchical approach had its immediate origins in the assumption by a disarmed samurai of the administrative role as leading bureaucrats enthused by the ideals of Confucianism. Under them, the artisan was vertically connected ultimately to the top through a series of hierarchical layers. At the same time the artisan acquired responsibilities and a mediated independence of varying degrees of potency at the village and community level, a situation which afforded opportunities for grass-roots industries. This stratified accommodation from central samurai to local artisan engendered the relative stability and trust which have been inherited right down to modern institutions. When industrialization overtook Japan, moreover, it struck so fast that the vertical social structure was in the main not superseded by a horizontal one, an eventuality which saw putative LMEs situated within a graduated structure with space to maneuver, but still overseen by the samurai. The Meiji Restoration was, therefore, implemented by reformist samurai who took over an economy already subject to considerable industrial and commercial development and in a state of gradualist transformation. They took responsibility for industrialization on the national scale.

The samurai could reach the country through established linkages, but success also depended on the willingness and capabilities of the artisan. In this process, Japan was distant enough from the advanced world to enable it to elaborate the so-called 'Meiji technology', a combination of the imported and the indigenous. This further stimulated the local entrepreneurial culture to supply the cities with all manner of old and new products. Underpinning this were many individual entrepreneurs, some of whom founded enterprises which still exist today as LMEs. Examples are presented of industrialization LMEs which adapted their indigenous technologies to the modern era or introduced new technologies into Japan. Koransha, on the other hand, is a case of a perennial LME which has retained its original identity while also bringing in the new.

6
Organization and Technology: 1912–1930

Overview

A crucial point about Koransha and the other firms briefly discussed in the final section of Chapter 5 is that, having been established as modern corporations some time during the period from the last quarter of the 19th century and the first decade of the 20th, they are still up and running as viable LMEs having made it to the 21st century. This infers a lot about the organization of the industrial structure in Japan and the arrangement of firms as discrete performers within that structure, as well as about how technology is dispersed and how it is acted upon and by whom. The shape of this organization and the implications of technology dispersal were to take on clearer definition in the second period, that following the end of the Meiji era in 1912 through to the culmination of the Second World War. So we now turn to how the samurai and the artisan confronted and contributed in their respective ways to this new phase, and then to the extent to which they merged their efforts, albeit haltingly and imperfectly, to meet the common challenges. Emerging from this activity, it will be argued, is a further substantiating stage in the evolution of the Japanese LME. This chapter explores the unfurling of events to the start of the 1930s, while the following chapter looks at the charged atmosphere of the nation on a war footing up to 1945 and the further implications this had for the industrial structure and the LME within that structure.

The samurai – bureaucracy and *zaibatsu*

So far we have pictured the samurai in his administrative role inherited from pre-Meiji times. During Meiji the character of the samurai started to change in two ways. First, meritocracy as expressed in academic ability started to take precedence over pedigree with the result that the bureaucracy at both national and local levels was drawn from a broader pool of the citizenry. Second, given the exigencies of the modern industrial state, the samurai was extended to encompass a business elite. Both of these samurai were to exert their influence on the shaping of the nation's industrial structure during this period from the end of Meiji to the beginning of the 1930s in ways that were to prove conducive to the gestation of the specialist LME.

The bureaucratic samurai and embryonic policy

As we have seen, throughout Meiji the bureaucratic samurai's overriding concern was to see established and operating the essential industries and organizations of the modern state. As this filled out by the end of that era and the larger business groups proved capable of running themselves, however, the administration started to direct more of its attention to guiding the activities of small business. By way of illustration, in the recession that followed the end of the First World War there was a resurgence of problems of excessive competition and mass production of inferior products, while at the same time there was a crying need for reliable exports because of a deteriorating trade balance. The bureaucratic response, starting in the 1920s, and aimed at assisting the cash-starved, although often exporting, smaller manufacturer was to reform the financial system, to instigate cooperative organizational structures through which smaller firms could operate, and to take its first tentative steps into the realm of what was to become 'industrial policy'.

However – and this will be a recurring theme in what follows – government essays at planning are often more interesting for their *unintended* rather than their intended outcomes, as the following example suggests. Increasingly concerned about the need for military preparedness by the latter half of the 1920s, the government embarked on a detailed examination of the nation's industrial structure, during the course of which it was soon made aware of the relative vulnerability of the machine tool industry. This was the start of a long courtship on the government's part lasting into the 1960s. It is the story of gifts dutifully received by the industry as deemed convenient, but its ultimate rejection of marriage to an industrial policy which sought to organize machine tool manufacturing on bureaucratically conceived lines, namely by having companies merge and by allocating production. In the event and speaking overall, amalgamation did not transpire; individual firms retained their autonomy and produced what they wanted.

The machine tool industry in Japan can in fact, without exaggeration, be described as an LME industry because many of the key companies thus engaged can be characterized as LMEs. And the deft thrust and parry of the industry and the firms within it as they notched up points while not giving an inch is in a sense an epitome of the rites of passage of the LME. Government intrusiveness strengthened the sinews of self-determination. The friction of asymmetrical agendas chiseled away at an emerging form on the corporate landscape.

The business samurai and its needs

Another contributor to this sculpting was the large business sector. As noted earlier, the initial preoccupation of the samurai bureaucrats was to create the semblance of modernity, superficial though it may have been, in order to contain as far as possible foreign incursion. The strategy, evolving haltingly and speculatively at first, was eventually to reach dry land as it moved with mounting confidence from semblance to reality. The agents co-opted by the central government to collaborate in the orchestration of this transformation were the business 'samurai', referred to as such here because of their *national* prominence and, most importantly emanating from that, their predilection and capacity for large-scale, intricate organization. Pioneering this function as agents were the large, diversified conglomerates known as the *zaibatsu*, the four major *zaibatsu* in that era being Mitsui, Sumitomo, Mitsubishi, and Yasuda. While their origins derived from non-manufacturing activities, they added them as time went on. Sumitomo demonstrates this, bearing in mind that some of the company names are not the same now as when they were originally incorporated and that in some cases the actual initial founding of the operation goes back further than incorporation as joint stock companies. Having established Sumitomo Electric Industries and Sumitomo Pipe & Tube in 1911 right at the end of Meiji, this *zaibatsu* went on to set up Sumitomo Rubber Industries in 1917, Sumitomo Chemical in 1925, and Sumitomo Heavy Industries in 1934 (JCH, 2004). On top of this, by the 1920s and even more so in the 1930s, the 'old' *zaibatsu* were increasingly joined by a breed of 'new' *zaibatsu*, as represented by the Nissan *zaibatsu* for instance, which embraced Hitachi among others, and by contrast specifically committed to well-defined areas of manufacturing such as electrical equipment, machinery and vehicles.

The First World War had caused shortages in many manufactured goods hitherto imported and had thus given an enormous fillip to domestic production. It had provided a springboard for the machinery, machine tool, metal and chemical industries, all demanding greater discipline, job commitment, skill and knowledge of a greater proportion of those employed in them than textile manufacturing, for instance. Given the speed of events, such people were not thick on the ground. It became incumbent upon manufacturers intent on retaining their competitive edge to recruit people with the potential for sustaining the company into the future. Not only had the company to train these people, moreover, it had to devise measures to induce them to stay. Thus were the seeds of guaranteed, life-time employment sown from the start of the 1920s by the *zaibatsu* and other large enterprises. Within their sphere of manufacturing, then, the large enterprises ostensibly had the best of everything – in terms of assets, technical expertise, qualified personnel, and business access. Organization was on their side; in fact, it was to a considerable extent of their own making.

It is also a fact that these large enterprises, not least because of the very form they took, coupled with the circumstances within which this transpired, contributed in no small way to the emergence of the LME. As manufacturers of raw and semi-finished materials, for one, they provided the smaller firms with the means to assemble final products. The smaller firms were thus partners in a complementary exercise through which the more capable, or more opportunely placed, could develop and grow. The large enterprises also used small manufacturers as sub-contractors, in so doing inducing the latter to modernize, while imposing their own technical standards on them. During the 1920s and 1930s, therefore, along with the central government, large firms increasingly became the essential channel for upgrading the technological capabilities of smaller firms as they redirected them from simply adapting traditional techniques to, instead, adopting processes geared specifically to modern industry (Morris-Suzuki, 1994:108, 129).

The artisan responding

Progressively more abstract though 'artisan' may become as we trace this historical phase from 1912 through to 1945, it nevertheless retains an explanatory potency in encapsulating the continuity and evolution of the reverse side of the coin – the non-samurai economic arena. Moreover, the artisan is the root cause of the development of the LME, most – although not all – of which started as small firms. And although central government and big business had their part to play in fashioning firms which were to become LMEs, the vast majority of small establishments from which emanated this select cohort were not the brain-child of the samurai. They were that of the artisan. Government and big business subsequently courted those smaller firms who showed promise and capability in the specific specializations in vogue at any given time, and indeed labored to upgrade the performance of small manufacturers in general, but these endeavors were ultimately founded on the aspirations and initiatives of a host of individuals who had set up shop in the first place.

A community base

To say that the artisan was the primary initiator of many a small manufacturing operation, moreover, implies that he was imbued with a sense of scope and maneuver. Within prescribed circumstances, the artisan was free to act. He was in his own territory where he was involved up front and shared responsibility for decisions taken. By the time we catch up with him again just prior to World War I, he had witnessed over 40 years of resolute but gradualist development. The small manufacturing venture of early industrializing Japan bore a number of characteristics, markings of which are still apparent to this day. First, the relatively slow shift to machinery had the effect of conversely preserving traditional handicrafts and techniques so that they, together with the minute attention to detail and precision which often went with them, were eventually woven into the modernized processes subsequently adopted. Second, these enterprises remained local, if not parochial, not least to ensure their supply of non-migrating workers. Third, given this spread across local communities throughout much of the nation, the best use was made of manpower which in turn meant that this countless array of small enterprises proved to be an immensely significant contributor to the early industrialization of Japan, accounting for well over half the country's manufacturing output.

By 1914, Japan had undergone a metamorphosis of no small magnitude. The artisan was well on his way to becoming the entrepreneur, just as the samurai was becoming the technocrat. The erstwhile embrace of the isolated rural community, now having been linked to a larger world by the railway and perhaps rudimentary telecommunications, had been superseded by the local authority, which nevertheless, and despite the more persistently pervasive intrusions of the modern state, had its eyes fixed first and foremost on the economic advancement of its own bailiwick. The guilds had been replaced by trade associations, or more specifically at the local level, manufacturing cooperatives (*kogyo kumiai*), tending to the needs of small and medium-sized enterprises in their respective sectors. The entrepreneur, the SME, the cooperative and the local authority were central among the political actors seeking to develop *and preserve* district or regional society and the economy supporting it. They comprised the new institutional framework whose opinions, beliefs and inclinations, with their homespun variations across the nation, commanded respect and attention even as they confronted forces of change which could not be gainsaid. Occupying a terrain conjoined to but contradistinctive from the central authority, they pronounced their inalienable presence and in so doing epitomized and embodied the artisan responding.

Modern technology and modernizing choices

Moreover, by the mid-1910s the machine had been infiltrating the artisan's workplace for a decade or so, contributing as it did to shaping a national economic structure which would begin to assume definite, discernible outline by the 1920s. For while the national enterprises were erecting their large-scale heavy and chemical industries, the smaller entrepreneurs – both rural and urban – were applying capital accumulated through sales in the domestic market and exports to investing in upgraded machinery and equipment to further develop their small-scale concerns in light industry ventures. For the nascent entrepreneur, then, the machine had become a partner in the practice of *monozukuri*, of making things. Furthermore, the machine was soon to be joined to electricity, bypassing the need for bulky steam power and thereby prolonging the *raison d'être* of many a tiny workshop. Hence, by the second decade of the 20th century, milling around the foothills of an industrial hierarchy crowned by a handful of *zaibatsu* conglomerates were thousands of mainly small, and often very small, manufacturing entities. In fact, by the end of that decade no less than 70 percent of those engaged in manufacturing were working in firms having fewer than 50 employees (Calder, 1988:145). More to the point, unsophisticated technologically, managerially and financially though the majority of these small firms may have been, they were nevertheless responsible for *most* of the nation's manufacturing; markedly more than in the West, the specific gravity of Japan's industrialization, and manufacturing in particular, resided in the smaller organization.

What is more, the larger Japanese manufacturer was not so large when contrasted with its Western counterpart. By comparison it lacked technological and capital resources as well as markets. So Fordist-type mass production was beyond the capabilities of practically all sectors of the Japanese economy, and even if it had been viable physically speaking, domestic demand was stagnant and export potential unpredictable (Tsutsui, 1998). To minimize risk and keep afloat in an uncertain environment, therefore, required alternative approaches. One of these was inter-firm coordination, as opposed to in-house investment and internalization by the larger firm, and this offered to the smaller manufacturer of parts and components the chance of survival as a supplier.

The artisan-turned-entrepreneur, in this scenario, with his machinery and newly acquired skills was given an extended lease on life, therefore. However, excessive attention to the relationships between large and small firms in a duality of mutual convenience, to client and subcontractor, and to the organizationally superior and inferior threatens to obscure the full picture. Japan was not simply an industrial juggernaut in the making; it was a society on the move. It was a society modernizing and in so doing taking in and adapting a cornucopia of novelties and concepts of living. As long as the samurai planner admitted the limitations of his range of intervention, which he implicitly did in industrializing Japan but which his Soviet communist counterpart in the 1920s and thereafter essentially did not, the entrepreneur enjoyed leeway to indulge his inclinations and ambitions. These were many and varied, concerning which bureaucratic state-building machinations were often not even remotely detectable. And yet they are grist to our story because, as will be seen from the accounts of a set of such entrepreneurs to be related shortly, not a few of these inspirations eventually culminated in LMEs which are now globalizing and therefore contributing in no small way to the variegated composition of Japan's presence on the international stage.

LME founding entrepreneurs

Many Japanese companies indicate when they were founded as distinct from when they were established as a corporation. Standing at the vantage point of around 1930, where this chapter ends and the next chapter begins, is as if being on a watershed from which we can look backwards and forwards. Looking back we see the formative phases of many firms on their way to becoming LMEs as they gravitate towards becoming incorporated. Looking forward we see a growing number of them established and assuming the persona of professionally managed corporations. In other words, during this period the validity of Nakamura's observation is already apparent. However, this does not necessarily mean that the family presence died out; in fact, in not a few cases of companies coming into their own during the period from 1912 to 1930 intense family involvement remains intact even now.

A variety of founders

The point to be made though is that the majority of what are deemed here to be non-family LMEs were at one time, needless to say, family LMEs, or at least were founded by an individual or group of individuals credited with starting the venture. In some cases their names are still emblazoned outside the company headquarters: the leading confectioner Ezaki Glico was founded by Ezaki Riichi in 1919; Hisamitsu Pharmaceutical, based on the southern island of Kyushu, had its origins initially under the name of Komatsuya started by Hisamitsu Nihira way back in 1847 – the year Thomas Edison was born as the company informatively states in its website history. But more often the company name gives nothing away and it is up to us to discover that, for example, it was Inoue Genzaburo who inaugurated Tomoegawa Paper in 1914 or that it was Oda Genzo who was the instigator of Mitsuboshi Belting in 1919.

On the family LME side, some enterprises took a considerable time to move from foundation to establishment, and one or two a very long time indeed. Takemoto Oil & Fat, for example, was already 310 years old when it finally made the move in 1945. There again, the truly venerable Fukuda Metal Foil & Powder, having been inaugurated by Fukuda Benseki in the district of Muromachi located in the imperial capital of Kyoto in 1700 to deal in gold and silver foil and powder, at last made it to the altar, one is tempted to say, in 1935. It is interesting to note also that, as of 2004, both these companies remained unlisted, at the time headed by, respectively, a Takemoto as president and a Fukuda as chairman (MKB, 2004:369, 446).

Moreover, the entrepreneurs who founded firms which blossomed over the first 30 years of the 20th century came in many different stripes and in their personal makeup and objectives reflected the rapidly changing environment in which they performed. True, the artisan in his pure form as portrayed in these pages was very much part of the act. Dutifully complying with this image, to give one illustration, were the three craftsmen who founded Dai-Ichi Kogyo Seiyaku as a manufacturer of surfactants and soap for the textile industry in 1909. But to go to the other extreme, it was none other than Baron Morimura Ichizaemon, an exalted member of the peerage newly constituted in compliance with notions of progress at the beginning of the Meiji era, who with his brother was the co-founder of Noritake, another manufacturer of refined ceramic tableware, in 1904. However, it is in the quality, training and accumulated experience of the unfolding breed of entrepreneur over the period that evolution is most striking; modern education, expertise and study abroad were remolding the creator.

Thus did Tanaka Gentaro, shortly after his graduation from Osaka Commercial High School, start a business in his own name in 1908 – representing as the sole agent an American maker of lubricating oil – which in due course was to become Teikoku Piston Ring. A decade or so earlier even, it was the holder of doctorate, Tanahashi Tonagoro, who founded the forerunner of Nippon Chemical Industrial in the heart of Tokyo to produce inorganic chemicals. Sakata Kyugoro was already an accomplished engineer in Hiroshima when he was introduced to the fountain pen by a British sailor acquaintance and so was capable of appreciating and emulating its design and functioning, to the point that he immediately founded a company named, with due deference to his communicator, Sailor Pen. When Tsuneto Noritaka founded Rasa Industries in 1913 he was currently the first director of the Fertilizer and Minerals Research Institute at the Ministry of Agriculture and Commerce – in the samurai camp no less. So he was well positioned to know that the island popularly referred to as Rasa (but officially as Okidaito-jima) in the Pacific some 400 kilometers southeast of Okinawa was rich in phosphate ore, a raw material for making chemical fertilizers, the exploitation of which was the company's initial business. And so on into the 1930s when Goto Yasutaro established what was later to become Origin Electric to produce in Japan the salt bath electric furnace he had studied with practical training in the United States. However, lists like this can only take us so far. What was the individual entrepreneur who jump-started such enterprises which contributed to the formation of Japan's industrializing economy as it progressed through these early years really like? Obviously impossible to generalize, but in the following we take six examples in an attempt to get more of an inkling.

Six early LME founders

The companies in Table 6.1 are listed in order according to date of founding. But we shall take them in a different order. First there are the three young men who initiated their own companies early on in life to produce more or less directly for the individual consumer: Mizuno Rihachi, Yamazaki Minejiro and Shimano Shozaburo. Then there are the two inheritors who transformed the traditional com-

Table 6.1 Six Entrepreneurs and the LMEs they Founded

Founder	LME	Details
Tanaka Umekichi	Tanaka Kikinzoku Kogyo	Year founded: 1885 Founded as: Tanaka Shoten Year incorporated: 1918 Business: Precious metals
Mizuno Rihachi	Mizuno	Year founded: 1906 Founded as: Mizuno Brothers & Company Year incorporated: 1923 Business: Sporting goods
Ashida Junosuke	General	Year founded: 1914 Founded as: Toyo Fukusha-Shi Year incorporated: 1940 Business: Printing products
Shofu Katei III	Shofu	Year founded: 1919 Founded as: Shofu Toshi Kenkyujo Year incorporated: 1922 Business: Dental materials
Shimano Shozaburo	Shimano	Year founded: 1921 Founded as: Shimano Ironworks Year incorporated: 1940 Business: Bicycle parts/fishing tackle
Yamazaki Minejiro	S&B Foods	Year founded: 1923 Founded as: Higashiya Year incorporated: 1940 Business: Instant seasonings

panies they took over to satisfy modern markets of the industrialized age: Ashida Junosuke and Shofu Katei. Finally, there is the Tanaka Umekichi, who, although the oldest of the group, set in motion a firm which became largely engaged in supplying intermediary materials to industry. In this Tanaka Kikinzoku represented the essence of the industrialization process and its attendant specializations, and it is the type of firm we shall be contemplating again in the following chapter.

Mizuno Rihachi of Mizuno

Mizuno is the true Horatio Alger hero of the bunch: in 1885, one year after his birth in Gifu Prefecture, his older brother died; eight years later when he was nine his father died. As recorded many years later in his memorial as a baseball hall of famer, his alma mater was Koubun Higher Elementary School, which he left at age 11 to restore the Mizuno family fortunes. By the time he was 13 his diligence had landed him a position in charge of procurement for a textile wholesaler in Kyoto, and it was during this time that he also acquired a passion for baseball and sports in general. Having parted from the company over managerial differences and then served on the front lines in the Russo-Japanese War of 1904, he founded, in the heart of the city of Osaka, Mizuno Brothers & Company comprising just two people, himself and his younger brother. He commenced with sundry items but soon zeroed in on sports goods. His timing was right because the Japanese public was beginning to take considerable interest in sports introduced from the West. He also showed a great talent for promoting his business by socializing with sports luminaries and organizing baseball and tennis tournaments. Having started with sales he moved into manufacturing, building a large factory in Osaka in the mid-1920s. One of the reasons purportedly for this initial venture into manufacturing was his dissatisfaction with the quality of imports from the United States.

This insistence on excellence was to raise Mizuno's game, but not before a new round of adversity. During the Second World War his main domestic factories were destroyed and he lost all his assets in Shanghai and Manchuria, as well as seeing the cancellation of the Tokyo Olympic Games scheduled for 1940. However, the Horatio Alger spirit was stubbornly undeterred: he had production running at one factory just six days after the end of hostilities and sales in motion four days after that. But this apart, what interests us at least as much in looking at what was by then Mizuno Corporation as a budding LME was the constant pursuit of reputation and quality, rather than short-term profit or growth for growth's sake. It was the desire to preserve the company's reputation that had Mizuno refuse to procure raw materials on the black market in the desperate days immediately after the war; it was the obsession with quality which made him resist heady expansion which he could have easily fallen for following the success of the Tokyo Olympic Games that actually did take place in 1964. Such attentiveness is key to the company now counting itself among the four major sporting goods manufacturers in the world and is epitomized in the Mizuno spikes Carl Lewis wore to set a new 100 meter sprint world record of 9.86 seconds in 1991, 21 years after the demise of the octogenarian founder in 1970 (IBO, 2005a).

Yamazaki Minejiro of S&B Foods

Just as Mizuno Rihachi had moved from a parochial birthplace eventually to Osaka, Yamazaki Minejiro likewise found his way to the big smoke, in his case from the village of Kanasugi in Saitama prefecture adjoining Tokyo, when he was just 17. Very soon after starting to work for a manufacturer there, like many a young rustic enthralled by the novelties of city life, he fell in love. His newfound object of devotion was curry and he immediately set about discovering how to concoct the essential ingredient, curry powder. Not that he would be the first in Japan to do so; that accolade apparently goes to a company now called Hachi Shokuhin, which came out with its version as early as 1905 (www.hachi-shokuhin.cp.jp). But having burned the midnight oil during a couple of years of arduous experimentation, he finally succeeded in manufacturing what he laid claim to be a purely Japanese version of this powder and on 5th April 1923, two months shy of his 20th birthday, launched Higashiya to produce and market it. Thus commenced a life-long commitment. As Mizuno communicated the joys of sport to the nation, so Yamazaki spread the word of spices. What they also had in common – as indeed do all six of our entrepreneurs in their various ways – was that they were agents in bringing aspects of an emerging universality of the early 20th century, or what we now call globalization, to the Japanese household, just as by the end of that century the companies they founded would be reversing the flow as they contributed to globalization in their distinctly Japanese fashion.

Yamazaki's particular forte was the leadership he displayed beyond the confines of his own company in campaigning for the industry's interests in Japan. This began with his chairmanship of the Tokyo Metropolitan Association of Curry & Sauce Manufacturers as of March 1936, to be followed by his inauguration of the Japan Pepper and Milling Industry Association in June 1941, and then immediately after the war the Japan Curry Association to consolidate extremely scarce supplies of materials. As well as being chairman of the latter, his post-war assignments included a stint as executive director of the Japan Curry Cooperative Association. For all this and contributions to society in general, Yamazaki was draped in honors, among which were the Medal with Blue Ribbon for his work for the curry industry, the Medal with Purple Ribbon for developing spices notably garlic powder, the Gold Rays with Neck Ribbon for 50 years of research into spices, and the

Order of the Rising Sun. Here was an artisan, an entrepreneur and an LME founder who relished public acclamation which, to be fair, he had substantially merited (www.sbfoods.co.jp).

Shimano Shozaburo of Shimano

Mizuno was a businessman who was a stickler for quality initially of items outsourced and subsequently of those produced in-house. Yamazaki was a pioneer who remained all his life closely involved with the spices his company developed and manufactured. But for sheer hands-on consistency Shimano Shozaburo is a very hard act to follow. A clue as to why this is so is that at the outset he was apprenticed to a trade in the city where he was born in 1894, Sakai, which is adjacent to Osaka and famous for its Japanese-style kitchen knives. Unlike Mizuno and Yamazaki making their way to the urban sprawl, Shimano was already in it and manifestly part of it from the word go. He became a live-in apprentice to a cutlery blacksmith there at the age of 15, then did the rounds of various ironworks in the vicinity before starting his own small factory at age 26 with a friend, still in Sakai where Shimano Inc. remains headquartered to this day. By 1923, two years after commencing, they had a complement of six employees working on four lathes, a milling machine and a drilling machine making freewheels, a gear for bicycles, as a subcontractor for Sakai Bicycle. Not content with being pinned down by a single client, however, Shimano himself also endeavored to sell to other makers as well as to wholesalers.

However, there was a big discrepancy in standards between the domestic item and imports. So from the middle of that year Shimano Shozaburo returned full-time to the factory where he concentrated all his energies on research and development to ensure product quality and uniformity. He improved the method of hardening the freewheels and he introduced a sandblasting technique to increase their luster. In addition to the product itself, he paid constant attention to upgrading the machinery, while bringing in a number of new machines of his own conception. Such successive technological innovation meant that by 1939 Shimano Ironworks was producing each month 100,000 freewheels with their 3-3-3 trademark and had captured 60 percent of the domestic market. Exports were also picking up, taking in China, Korea and Southeast Asia, but also extending as far as India and Africa. Even during the difficult five years or so immediately after the war when the necessary materials were very hard to come by, Shimano kept his hand in, breaking technological ground with a new method of hot forging and a further advance in the freewheel hardening process. And just one year before his untimely death from illness in 1958 he was absorbed in making technological improvements to a copy of an internal hub gear mechanism originating from Britain – there was still catching up to do – and investing in equipment with the aim of building a system to increase the production of these gears from 3,000 to 50,000 units a month. The hands-on man had toiled on the shop floor for decades to virtually the point, which he would not witness, when his company would take off big time in the 1960s under the inherited direction of his three sons. Nearly five decades later, American Lance Armstrong would ride to his seventh consecutive victory in the Tour de France on a bicycle fitted with Shimano Dura-Ace components especially made for road-racing (IBO, 2005b).

Ashida Junosuke of General

The founders of General and Shofu, Ashida Junosuke and Shofu Katei respectively, were also sons, but adopted ones through marriage in the Japanese tradition of preserving the household, or *ie*, as both a social and economic entity. They were both inheritors of established concerns who applied their entrepreneurial initiative in combining native and imported intelligence and techniques to transform what they took over into modern businesses in response to the evolving demands of an industrializing nation. Being an adopted heir also entailed assuming the surname, at least, of the head of the household, which is what Ashida Junosuke did when he became the son-in-law of Ashida Rihei in 1906, having done duty, like Mizuno Rihachi, in the Russo-Japanese War. Father-in-law Rihei might have been traditionally inclined as far as the way he chose to keep his business going was concerned, and indeed in the type of business he ran, but in the actual *practice* of business he was ahead of his time. Having been, like Shimano Shozaburo a few decades later, a live-in apprentice, Rihei set up his own stationary outlet for selling *fude* (writing brushes) and *sumi* (Japanese ink), which were essential items for any kind of business or trade at the dawn of the Meiji Restoration for entering accounts in the traditional ledger. Where he was different was in, what would now be called, his marketing approach. He brought together skilled craftsmen from nearby Arima, famous in the art of making *fude*, to produce high-class items which he then sold under his own brand name exclusively available from his shop. What is more, he did not adhere to the conventional practice of sitting around waiting for business to come to him, nor did he go through wholesalers, but instead circulated his trading-house and banking clients in downtown Osaka to promote sales directly.

This go-getting aggressiveness evidently rubbed off on his son-in-law. At the time Junosuke assumed duties as the heir to the business, Japan, having defeated Russia, was joining the ranks of the Great Powers and the country resonated with the tempo of modernization. For the stationary business this translated into a shift in demand from *fude* and *sumi* to Western pen and ink. For the house of Ashida Junosuke it heralded an expansion into a complete range of office equipment. At the same time, Osaka was in the process of becoming the largest commercial centre in the Far East, so that Junosuke, with his reputation for fastidiousness, saw his customer base steadily substantiated. Western ink was the big talking point at the time. But the more Junosuke became preoccupied in producing it, the more he became convinced that it was not the ink as such which held the real future in office equipment but the carbon copying paper coated with that ink. This it was, therefore, that he concentrated on developing and manufacturing, and it was his belief in the long-term potential of carbon paper that induced him to found Toyo Fukusha-Shi Limited Partnership, the forerunner of General, in 1914.

The company started in the production of carbon paper by coating an ink, of a constituency contrived by Junosuke himself, on a fine-quality white Japanese paper traditionally made in Kawanoe in nearby Aichi Prefecture, thereby presenting a classic illustration of the melding of the old and the new still prevalent in that era. In those days it was the custom for wholesalers (*tonya*) to buy in from the manufacturers, ascribe various brand names to the product and sell on to the customer. But having established his own 'Mikado' brand, Ashida Junosuke then took another page out of his father-in-law's book. Confident of the quality of his product, he refused the option of remaining as a cheap subcontracting processor to wholesalers, but rather elected the tougher and more precarious route of selling direct. General, the successor to this approach, retains its aggressive nature to this day, boasting profitable plants in the United States, Britain, China and South Korea among others. But we leave it when it was still Toyo Fukusha-Shi and a tale of teething problems of the 'latecomer' and how they were resolved. Having initially relied almost entirely on manual operations, in the 1920s the company introduced a coating machine from Germany. However, nobody knew how to use it, so it stood idle in a corner for quite a number of years. It was only resurrected when Junosuke's third son, armed with the requisite know-how having graduated in science from Kanazawa Specialist High School, joined the company in 1933. That was just one year before Junosuke's death from acute pneumonia at only 55 years old while on a business trip to Tokyo (www.general-jp.com).

Shofu Katei III of Shofu

Shofu, located in Kyoto, has some similarities with Koransha, the ceramics and insulator manufacturer described in Chapter 5. For one they have both had a succession of leaders, bearing the name of the originator of their business establishment, taking them through to modern times. In Shofu's case this went as far as Shofu Katei V who chaired the company up to the new millennium. Shofu Katei III was born Inoue Jotaro in 1870 and as a young man was apprenticed to a second-generation ceramics artisan, Shofu Katei II. His outstanding work and potential were rewarded by being made heir apparent and by marriage to the boss's daughter. In 1906 he launched his own business with the intention of manufacturing insulators like Koransha, although in his case they were for export. It is worth noting also that the source of native input was likewise local: while for Koransha it had been Arita-yaki, for Shofu Katei it was the traditional *Kyoto* ceramic art of Kiyomizu-yaki. Evidently insulators were not to be Shofu Katei's destiny, nor exports for that matter, because he was soon taken by the attractive proposition of providing a growing domestic market with porcelain teeth, all of which were imported at the time. Dentures were not unknown in Japan, having been around throughout the Tokugawa period. But they were expensive works of art and mostly made of wood. So here was another example of how an item already existing within Japanese society could be enormously improved by the infusion of Western technology.

For such a venture Shofu Katei was well suited, combining as he did the dual talents of entrepreneur and researcher. Eventually he founded Shofu Toshi Kenkyujo, or the Shofu Porcelain Teeth Research Institute, and started to develop porcelain teeth to meet the particular requirements of the Japanese. In other words, there was sufficient native expertise and knowhow on tap to adapt the product to local preferences rather than merely accept unquestioningly the imported item as the model. Through his Institute, Shofu Katei could call on Japanese specialists in the anatomical structure of teeth to collaborate in this endeavour, the culmination of which was the 'Shofu Anatorm Form' which was to constitute the basis of the denture-making technology of the company established in 1922, initially as Shofu Dental Mfg. Co., Ltd. From the outset it was a modern company in the sense that it was established with a defined core competence resulting from a deliberate program of research and development, the consistent practice of which has ever since been a defining characteristic of the company, as symbolized by the building of the Shofu Institute as early as 1936. Four years before that, in 1932, national recognition had been afforded to Shofu Katei III for his success in being the first manufacturer of porcelain dentures in Japan. For it was then that the Ministry of Home Affairs inaugurated the Pharmaceutical Promotion Committee, appointing him as the person responsible for dental matters. (Evans, 2003; www.shofu.co.jp).

Tanaka Umekichi of Tanaka Kikinzoku

Ceramics comprise a material fashioned from clay. Ceramic products we have mentioned – tableware, insulators and dentures – are easily recognized for what they are. But a lot of what is extracted from the earth is not so easily recognized once having been refined into materials and then processed into products. What is more, many of these materials and products are then incorporated into intermediary and finished goods where they remain to all intents and purposes invisible to the layman. This is important as far as we are concerned because, although it is evident from the above biographical snippets that LMEs can be the manufacturers of end-user goods, many other LMEs are partially or wholly intermediary suppliers and therefore far less likely than the former to attract public awareness. Tanaka Kikinzoku, also rendered as Tanaka Precious Metals, is one such company and platinum, a rare white heavy metallic element extracted from the earth, is at the core of its production activities. In fact the company starting out as Tanaka Shoten had its origins in its founder's dealings in gold and silver coins. Tanaka Umekichi commenced his business by collecting small coins from shrines and temples, which he handed over to banks in exchange for notes to be transferred back to the shrines and temples for a commission. Soon he was melting down American dollar coins for bullion. This business was to continue and grow for decades as demand increased for bullion to meet Japan's shortfall in supply to finance its trade with the West. As an exchange house, Tanaka's company bought dollar coins and separated them into their constituent metals to get gold and silver. Demand became so intense in fact that the dollar coins were separated and refined overnight for sale the next morning.

But by the time Tanaka Kikinzoku was established in 1918, under the chairmanship of Umekichi, the realization had taken hold within the company that the order of priority was platinum/gold/silver, not gold/silver/platinum as was generally presumed at the time. Not that the other two metals were downplayed. Tanaka Kikinzoku is after all today the world's largest supplier of bonding wire, much of it incorporating gold or gold alloy, which is used for a wide range of products including ICs, LSIs and transistors. To be more exact, what is entailed here are platinum group metals (PGMs) comprising platinum itself together with allied palladium, rhodium, ruthenium, osmium and iridium, all of which Tanaka Kikinzoku now handles. And placing platinum on the top pedestal was a momentous insight at that early stage, especially when one considers that, according to a recent estimate by the International Platinum Association, 'about one in four goods we use daily owes its existence to PGMs'. They turn up in everything from fountain pens to aircraft turbines, from anti-cancer drugs to mobile phones, from catalytic converters for automobiles to ceramic glazes (www.platinuminfo.net).

For Tanaka Shoten this all started when it became Japan's first manufacturer of platinum industrial products in 1889. The platinum itself was largely imported from Russia. But there was some osmiridium (osmium + iridium) in Japan's northernmost main island of Hokkaido which Umekichi and his chief engineer would purchase on trips up there, the company having devised the technology for refining it in 1907. Progress thereafter was sustained and deliberate, so that by the time of the chairman's demise in 1936, Tanaka Shoten and then Tanaka Kikinzoku had initiated an array of products, most of which owed their existence to PGMs: electric stem platinum fine wire in 1907, platinum electrodes for the manufacture of potassium chlorate for gunpowder in 1914, platinum-iridium hypodermic needles in 1915, gold-platinum spinnerets in 1925, platinum gauze for catalysts in 1930, platinum-rhodium alloy catalyst gauze in 1933. Today the company's clientele spans the industries of electronics, automobiles, telecommunications, ceramics, health and environment and more. In all probability the moneychanger taking his cut at a shrine in the mid-1880s had little idea that his ship would call at so many ports. But he started it, nevertheless, and steered it a fair distance himself (Tanaka Kikinzoku Kogyo K.K., 1985; Evans, 2003).

Summary

To this juncture the samurai has been depicted as a bureaucratic administrator. It was this administrator who laid the groundwork for the modern industrial state. However, the complexity of industrialization required that the administrator be supplemented by a complementary elite, referred to here as the business samurai. These two were instrumental in shaping Japan's industrial structure.

Having been mainly concerned with establishing the essential industries and institutions during the Meiji era, by its close the samurai bureaucrat was turning his attention to smaller firms to the point that by the 1920s he was starting to map out what came to be known as an 'industrial policy'. Emerging also were the business 'samurai', represented in the first instance by the 'old' zaibatsu centered on financial institutions and subsequently joined by the 'new' manufacturing zaibatsu. They also offered their own stimulus to the smaller firm as upstream suppliers, as well as subcontracting clients which contributed to the technological upgrading of the latter.

In the 40 years to 1914, artisans as entrepreneurs had initiated many businesses, having been imbued with a sense of scope and maneuver. They existed in communities nationwide with which they had to find some accommodation. By 1914, entrepreneur, SME, cooperative, and local authority were working together in a conscious community setting. The machine was by then a partner of *monozukuri*, soon to be joined by electricity which facilitated the survival of very small operations. There were many more firms in manufacturing proportionately than in the West, sustained by inter-firm coordination as against large-firm internalization. There was a wide variety of founders, some whose names still survive outside company headquarters; they included craftsmen, men with doctorates, and even peers of the realm. Six of them were: Tanaka Umekichi (precious metals), Mizuno Rihachi (sporting goods), Ashida Junosuke (printing products), Shofu Katei (dental materials), Shimano Shozaburo (bicycle parts), and Yamazaki Minejiro (instant seasonings).

7

War and the Molding of an Industrial Structure: 1930–1945

An industrial nation

When Tanaka Umekichi passed away in the mid-1930s, Japan had been launched on a course of transformation shaped both by what it had become over 60 odd years of industrialization and by where events and its own militaristic excesses were now taking it. There could be no turning back to isolation and an idealized economic stasis sustained by traditional agriculture and native crafts. Despite the increasingly shrill calls of nationalist extremists for the country's 'Japanization', the underlying mobilizing forces of change – even at the height of belli-cose fervor and patriotism – were simply going in a different direction, and largely in the direction the West had set at that. The sports goods, curry powder, bicycle gears, carbon paper, ceramic dentures and hypo-dermic needles encountered in the previous chapter were all evidence of that, as were the country's transportation system, its educational institutions, its police force and the plant operations and facilities of its major manufacturers, all of which owed their existence to a Western blueprint. Japan was becoming an industrial nation; by 1940 its sec-ondary sector would account for approaching 50 percent of its net domestic production as against just over 20 percent at the start of the 20th century, while over the same period its primary sector had declined from some 40 percent to under 20 percent (Okazaki and Okuno-Fujiwara, 1999).

War footing

However, this was not and has not even now led to wholesale conver-gence with Western – and more specifically Anglo-American – practices;

Japan has had its own singular path-dependent progression. As we have seen, in its case the 'samurai' and the 'artisan' worked in from opposite ends of the spectrum to effect the country's modernization. While the central government and large corporations imported advanced foreign technology, smaller firms, often with the assistance and encouragement of local authorities, busied themselves with upgrading and grafting technologies, parts of which could be traced to traditional knowhow, like General, Shofu and Tanaka, or bringing to the market items incidental to a closer engagement with the outside world, like Mizuno, S&B Foods and Shimano. In this process the traditional, often assuming mounting sophistication both technologically and organizationally, linked up with the modern, a phenomenon which was now to take on added pertinence as Japan tensed itself and attempted to marshal its assets in an atmosphere of growing hostilities.

As part of this endeavor, the Ministry of Commerce and Industry (MCI) sought to control and channel production by selecting and favoring manufacturers – small ones with a traditional provenance included – in an attempt to ensure ready supplies. How far it actually succeeded in its own objectives is open to debate, but the upshot, as we shall see, was a further delineation of Japan's industrial structure as more and more companies moved up to the front line and the scope for eventually achieving LME status was broadened.

An emerging structure

The Ministry of Commerce and Industry was actually established in 1929, this being the direct outcome of the government's intensifying concern about deficiencies in the industrial economy and the consequent need for planning, which had in the first instance led to the creation of the Resources Board (*Shigen Kyoku*) two years previously. What was to follow, from the central government's perspective at least, was a persistent and unerring tightening of the screws of direction and control until the denouement of military defeat in the summer of 1945. However, although as the 1930s progressed the prime *military* motive for this attempted organizational reordering and streamlining was doomed to ultimate disaster, elements and structures thereby transpiring in the process – both intentionally and not so intentionally – were to survive into the postwar era. Intentional were, for instance, the campaign to augment ministerial influence over the economy as a whole and, at the microeconomic level, the elevation of corporate management to virtually unassailable authority at the expense of ownership as represented by the shareholders. Enhanced efficiency to meet the exigencies of war purportedly warranted such institutional tinkering as events unfolded at the time, but the resulting infrastructural forms – like administrative power of intervention and executive supremacy over other corporate stakeholders – were still largely intact to fight the battles of economic recovery and catching up with the West for most of the rest of the 20th century, well after the war was over.

Arguably not nearly so intentional, on the other hand, was the scope afforded to more and more smaller firms to climb into the ring to test their mettle, many of them demonstrably specialized, with a growing number concentrating on narrowly defined core competences. These were the better SMEs needed to fill the gaps – to a considerable extent but not exclusively – as subcontractors for manufacturing jobs from the larger assemblers. However, while they were to be encouraged they were to know their place. Planning looked for order and predictability, not high-flying ambition. The factor unforeseen at the time, therefore, was the degree to which not a few of these enterprises were to contribute to the evolution of the Japanese economy thereafter, notably during the high-growth era from the late 1950s to the early 1970s. Japanese LMEs were also to have their place in the sun as constructive actors in the recovery, the catching up and, indeed, in not a few cases, the overtaking.

Scope for the latent LME

'Samurai' fiat, then, did not simply ordain the creation of the LME; the 'artisan' is not to be so easily gainsaid. As has already been made abundantly clear, throughout Japan's industrializing period many founders of smaller firms were entrepreneurs primarily preoccupied in developing business on their own account. True, they could be induced to change course, like the forerunner of Owari Precise did in 1938 from timepieces to screws for military aircraft, as their original markets disintegrated and new ones became self-evident – bordering on compulsory even. But the assumption that upon laws and directives being promulgated everyone falls into line is seriously flawed.

The center's limitations

Not least of the reasons for this was that, although the government bureaucracy, and within it most especially the Ministry of Commerce and Industry, occupied the citadel it was not the sole occupant. There were others contesting the turf, frequently at odds both with the government and with each other. The military – the Army and Navy – for one looked to further the advancement of the new *zaibatsu*, like Nissan, more sympathetic to its cause than the old *zaibatsu*, and for its supplies the military was in the habit of bypassing the control associations set in place by the bureaucracy. Even the final turn of events, the reconstituting of the MCI as the Munitions Ministry in 1943 in a last-ditch concentration of economic regulation under a single ministry, could not plug the gaps here. Then there were the old *zaibatsu*, not united even in the thick of war, whose unrelenting competition among each other for market share often toppled the agreements necessary for workable cartels, and who had ideas conflicting with those of the bureaucracy as to how the economy should be run. Finally, opposing the *zaibatsu*, even as they supplied them, were a host of smaller firms, seeking a more level playing field through financial reform, while receiving some support from the military due to lower-class affinities their members often shared in common.

Given the circumstances at the time, however, which after all must have evoked a common desire for Japan to win through in the long run, these elements had no choice, despite projecting nodes of incompatibility, but to coexist and collaborate in a patchwork of compromise. And it was this state of affairs, every bit as much as the designation of industries and companies in a program of mobilization for war, that provides additional insight into the particular makeup that the Japanese economic structure has assumed and the LME within it. For the inconsistencies of the 'samurai' message offered the 'artisan' scope for maneuver; what started out from the centre as a straight line of intention ended up curved and warped into compromise and contingency. One such bureaucratic straight line was the wish, in the interests of enhanced efficiency, to concentrate manufacturing either by having smaller companies combine into larger units or by eliminating them with the objective of centralizing production in fewer plants. But although some headway was made in mergers, in chemicals, textiles and food processing, for instance, materials shortages and bombs were infinitely more effective in forcing firms to close down than attempts at official suasion.

The primacy of demand ...

However, the single most important contributing factor in keeping smaller firms afloat and inducing the foundation of new ones – provided they were engaged in the right industries that is – was demand. Impetuous demand, and the response to it, dominated the supply routes, muscling aside and short-circuiting the more refined niceties of planning and order. The right industries, as the war atmosphere thickened, were heavy industry-related. Favored at the expense of light industries, especially after the enactment of two laws in 1937 restricting the production of non-essential goods, the steel, non-ferrous metals and machinery industries expanded as they benefited from the government's concentration of available resources upon them. The upshot was that by 1942 the heavy industry sector commanded 70 percent of total industrial production as against 50 percent in 1936 (Yoshihara, 1994:15). The results were manifest. It was estimated, for example, that by the end of the war Japan had twice as many facilities in workable condition for steel, machine tools and other machinery made from steel than at the time when it invaded Manchuria in 1931 (Tsuru, 1993:8).

Managerial and technological efficiency were the key objectives of the government in striving to keep up with this demand, but in actual fact specialization and a proliferation of entrants was the way much of this demand was met. Proliferation was particularly apparent with respect to the machinery industry, prioritized as it was by the government from the very start of hostilities with China in the early 1930s. From 1932 to 1938 the number of establishments with more than five employees and classified as machinery manufacturers increased no less than five-fold, and that does not include probably many more with five or fewer workers (Friedman, 1988:43). Just how minute the specialization could be is illustrated by the *byora* industry in Osaka and what is now the adjoining and separate city of Higashi-Osaka. *Byora* means fasteners, this being the generic term for nuts, bolts, screws and rivets, of which these days there are tens of thousands of types. The business had started to be concentrated in that area with the production of bolts by specialized manufacturers to supply the shipbuilding industry during the mid-Meiji period, notably to satisfy demand for Japan's two brief wars with China and Russia in 1895 and 1904 respectively. This afforded the impetus for expansion as an industrial district and the supplying of other customers including machinery makers who also settled in the vicinity, so that by the mid-1930s the long-established combined with the new arrivals, both often very small, constituted a pool of demand for the *byora* makers, likewise often very small, in a close-knit web of constant interaction (Taguchi, 2000:133–4). Given the structure that that industry was assuming in Japan at the time, much of the demand from the top ultimately had resort somewhere in the manufacturing chain to the specialized *byora* maker and the other specialists he collaborated with and serviced.

... and the rise of the specialist

Here, in fact, was a localized microcosm of the economy as it was evolving. For manufacturers overall, even those that grew large, pre-ferred to remain specialized in fairly narrowly defined areas of expertise rather than integrate vertically and diversify production. As an alter-native they sought cooperative relationships with other specialists who could supply them with components and parts, this practice being particularly prevalent in the automobile, electrical equipment and machine tool industries, for example. So, rather than internalization by the more powerful players, the prevailing trend was in the direction of integration of activities in subcontracting networks, practiced to the extent that coordination was practical and suppliers were capable of fulfilling specifications. This trend was not new, the approach having been pioneered decades earlier by the textile weaving industry, for instance. But the momentum was now intensifying primarily, it would seem, because of the rapidity of change firms were encountering from the advent of the 1930s on. This acceleration made the larger corpor-ations wary of taking on any undue extra risk inherent in absorbing added manpower and plant, and besides which capital for expansion was at a premium and markets were not substantial enough to justify meaningful economies of scale.

A symbiosis of need was the outcome. While the subcontractors embodied the complexity and expenditure which the larger assemblers refrained from taking on, conversely these same subcontractors also constituted the source of supply for orders the assemblers increasingly chose not to, or no longer had the capacity to, fully handle by them-selves. Thus did Toyota have some 400 suppliers around the start of the Pacific War, while at the end of it aircraft manufacturer Nakajima, by that time swamped with orders, had to depend on subcontractors for over 40 percent of its output. The subcontractors for their part could look to their clients for technological and financial assistance as well as broader business prospects than they themselves were in a position to develop. This had two important consequences as far as the gestation of LMEs in Japan was concerned. First, the interconnection across dif-ferent sectors of the economy helped nurture a diverse, flexible and creative technological base out of which many companies with high levels of proprietary expertise were to emerge. Second, the inter-firm integration as described above had the effect of constraining growth through in-house diversification – even if the inclination had been there – to the point that firms directed much more energy to product and process upgrading for superior added value. The result was that many of them remained relatively small but nevertheless specialized and appreciated for that, as well as being, to varying degrees admittedly, organizationally autonomous.

LMEs in the making

Not surprisingly it was industries beginning to come into their own which contributed most to the broader scope of participation by newcomers and others prepared to take on fresh challenges. The domestic automobile industry is a classic case in point. Although, it is worth noting in passing that the first car produced entirely in Japan was the 'Otomo', which was the magnum opus of Toyokawa Junya, another inveterate Japanese individual of the first decades of the 20th century. Disdaining to use anything but equipment and tools produced by his own company, Toyokawa manufactured and sold some 300 'Otomo' – even exporting to Shanghai – through the Taisho era (1912–1926), which happens more or less to correspond with the historical phase discussed in the previous chapter (JMIF, 2002). However, by the 1920s the big boys were General Motors and Ford, with plants in Osaka and Yokohama respectively, assembling knockdown kits coming in from the United States. In the early 1930s, the main domestic makers, even lumped together, were producing only a few thousand units a year (Fruin, 1992:261).

Automobiles: a new domestic industry

But things were starting to change, so that by the end of the decade, assisted on the way by the enactment of the Automobile Industry Law in 1936 which promoted domestic manufacturers in the interests of military preparedness while effectively sidelining the Americans, familiar faces had emerged from the pack – Isuzu, Hino, Nissan and Toyota. And the point that has to be firmly underlined is that the emergence of these companies and their subsequent growth as they experimented with mass production techniques was, at total variance with Toyokawa's 'Otomo', heavily reliant on subcontracting. The two – mass production such as it was and subcontracting – in fact, moved in tandem from the outset. True, much of the subcontracting was in effect in-house, involving as it did companies within the group or closely affiliated to it. Thus did Toyota Motor co-opt the services of Aichi Steel, already the steel manufacturer for the group's parent, Toyoda Automatic Loom Works, while forging close – or what would now be referred to as *keiretsu* – alliances with the likes of lighting equipment manufacturer, Koito, and a maker of bearing metals,

Daido Metal. Nevertheless, considerable variety was apparent, not only in how these companies commenced their engagement with the automobile industry but also in what they subsequently became. Aichi Steel, for instance, remains to this day a stalwart member of the Toyota Group; Koito, on the other hand, while maintaining a close association with Toyota, is exploring greater autonomous expression by expanding into aircraft parts as well as through its operations abroad. There again, Daido Metal, which is an LME by our definition, now delivers to all the domestic automakers, as well as to shipbuilders, construction machinery makers and others.

The specialist moves in .... and remains

Reference to the companies listed in Table 7.1 affords further insights. First, it must be stressed – as it will be again concerning the inauguration of successful LMEs throughout the postwar period – that the individual founder is far from being drowned in a sea of bureaucratic corporate design. The founder of Nippon Piston Ring was Suzuki Tomo-

Table 7.1   Automobile Component Specialists Established between 1930 and 1945

Company	Established (Founded)	Location	Product Range Initial → Current
Nippon Valqua	1932 (1927)	Tokyo	Brake linings → industrial packings
Nippon Piston Ring	1934 (1931)	Saitama Pref.	Piston rings* → engine parts
Akebono Brake	1936 (1929)	Saitama Pref.	Woven linings and clutch facings → brake mechanisms
Meira	1936 (1932)	Nagoya	Bolts → fasteners
Pacific Industrial	1938 (1930)	Gifu Pref.	Valve cores for autos → auto and electronic parts
Imasen Electric	1939 (1932)	Aichi Pref.	Electric horns* → auto parts
Diamond Electric	1940 (1937)	Osaka	Ignition coils for autos → electronic control components

*Indicates that the company claims to be the first Japanese manufacturer to produce the given item in Japan.

nori, while the founder of Imasen – and hence the name of the company – was *Imai Senzaburo*, and both laid claim to being first off the starting blocks in Japan with their respective products, piston rings and electric horns for automobiles. Second, with the exception of Diamond Electric, the founding dates of the rest are clumped *before* domestic automobile industry really started to take off in the middle of the decade. Their initial clientele, therefore, spanned not only the small domestic makers but also the replacement-parts market, mainly for foreign makes. To that extent they were independent actors venturing into a growing industry, rather than merely pawns responding to the drum of domineering assemblers.

Third, from the word go, the automobile industry in its mass production era, has been a demanding and unsettling customer. One way to handle this has been to diversify risk by specializing within a fairly narrow band still within the automobile industry, like Nippon Piston Ring, Pacific Industries and Imasen. Another is to extend somewhat further into an adjoining specialization, while still keeping a firm grip on the original core competence, like Diamond Electric. Currently this company retains a 30 percent share of the domestic market for automotive engine ignition coils, although it is now principally a manufacturer of electronic control parts for air conditioners and water heaters. Yet another alternative is redefinition and a redirection of emphasis which was the option taken by Nippon Valqua. Having been founded to make brake linings in 1927, by the time it was incorporated five years later Nippon Valqua's principle commitment was to industrial packing where it has been lodged ever since. On the other hand, the story of Akebono Brake is one of incremental accumulation consistent with its origins. Having likewise started with brake linings, it remained strictly on course, ultimately progressing to brake units, disk brakes chiefly (Nakamura, 1990:202).

Note that none of these companies – including the incorporated Nippon Valqua – have strayed far, if at all, over some 70-odd years of existence into vertical diversification, and this observation brings us to the final point. As we have seen earlier in the case of Tanaka Kikinzoku with its precious metals, rather than diversifying vertically, such firms have shown a marked tendency to explore business opportunities *horizontally*. This is fairly self-evident as a strategy for Nippon Valqua, which now serves the packing needs of the electrical and electronics, automobile, vacuum, food, and aircraft industries, as well as being engaged in semi-conductor processing. Perhaps it is less so for Meira, though. Here we are back to the ostensibly mundane and unfashionable world of *byora* – fasteners – again. But, as already noted, the demand for them is ubiquitous. Hence, this LME, while retaining a steady consistency as a supplier of bolts to the automobile industry, has extended its business to the aerospace industry, and also – less self-evidently – to the medical industry. Meira's *byora* now find their way into the human body, and in so doing epitomize the elaboration of core competences typical of many an LME (www.meira.co.jp).

Machine tools: an 'LME industry'

Epitomizing in another sense the aspect of the LME is the machine tool industry. With respect to Japan this industry can be employed both to bring together the strands of the discussion to this juncture and to function as a link with the postwar era. Machine tools are power-driven devices for cutting, forming and shaping metals to produce new mechanical devices such as automobile components, electrical parts, weapon components and medical equipment. They are critical to manufacturing in that their performance determines the quality of the end product. The machine tool industry makes things to make things; it is at the very heart of *monozukuri*. Although it is only a small part of the industrial machinery industry, there are a large number of companies engaged in it, both in Japan where small manufacturing firms are numerous anyway, but also in the United States, for example, where they are less so.

The reason here for categorizing machine tools as an LME industry in Japan is that, although there are some large players as well as machine tool divisions in large corporations, and there are many small and very small participants also, the girth of the industry comprising most of the significant specialist makers is currently mid-sized. This outcome is not a little to do with the support the smaller participants have provided. For by collaborating with their medium-sized counterparts these smaller firms absolved the need for internalization, thereby contributing to the subsequent emergence of LMEs in the industry. And the reason for singling the machine tool industry out at this point in the discussion is that its domestic development was pivotal to Japan's progress during the period from 1912 to 1945, especially from when it was prioritized from the start of the 1930s, while its rapid postwar advance has symbolized Japan's impressive economic evolution over the past half century or so.

Some resilient pioneers

The firms given in Table 7.2 were all founded during the period from 1912 to 1945. Their accumulated experiences since their foundation

Table 7.2   Machine Tool LMEs Established within 1912–1945 Period

Company	Established (Founded)	Location	Product Range Initial → Current	Stock Market Section	Listed
OKK	1915	Osaka	Pumps/water metres → MCs/NC milling machines	1	1949
Okuma	1918 (1898)	Aichi Pref.	Noodle-making machines → MCs/NC lathes	1	1949
Okamoto Machine Tool	1935 (1926)	Yokohama	Grinding machines → surface grinders	2	1963
Koike Sanso	1936 (1918)	Tokyo	Gas cutting equipment → gas-powered machine tools	2	1969
Tsugami	1937 (1923)	Tokyo	Block guages → CNC lathes and MCs	1	1949
Hamai	1938 (1922)	Tokyo	Gear hobbing machines → hobbing and lapping machines	2	1963
Takisawa Machine Tool	1944 (1922)	Okayama	Lathes → NC lathes/MCs	2	1962

and subsequent incorporation to just beyond the turn of the millennium can be cited as condensed compendia, as it were, of what we have seen of the rites of passage of the LME and what we are about to see as it edges towards globalization. Condensed, that is, in the manner these machine tool LMEs have traversed time and space, as compact as golf balls, keeping true to course while in essence and identity changing hardly at all; five of them – Okuma, Okamoto, Koike, Hamai and Takisawa – still bearing the founder's name by way of symbolization. But condensed to boot because the machine tool industry, at least in Japan, has shown itself singularly unyielding in its commitment to strictly delineated core competences, as can be seen by the limited distances traveled over the years from the initial to current products. Okamoto and Koike virtually seem to have been running on the spot in that sense. Only Okuma came in from outside among the companies in this group, and that was very quickly, because it was already starting to manufacture machine tools in 1904, six years after its foundation, on its way to becoming Japan's top manufacturer in terms of value by 1937 (www.okuma.co.jp).

All the companies given here became listed on the Tokyo and other stock markets during the postwar period, three of them – OKK, Okuma and Tsugami – in 1949, immediately they were able to after the lifting of the ban on machine tool manufacturing by the Occupation forces. The other four took advantage of the easier terms for listing with the inauguration of the Second Section in the early 1960s. Nakamura (1990) regarded joining the stock exchange as a sign of maturity in corporate organization accompanying the phased separation of ownership and management of the LME. However, as with Lego among the universal LMEs in Chapter 3, there are significant examples in Japan, too, of firms choosing not to be listed, some of which we have already encountered, these being Koransha, the traditional and modern ceramics manufacturer, Tanaka Kikinzoku, the producer of precious metals, and Meira, the specialist in fasteners. That having been said, although joining the stock exchange has effectively been the pronouncement of enhanced expectations for many a firm, the approach of most Japanese LMEs – whether listed or not – when they have expressed such expectations through international activities is best described as measured, a fact well illustrated by the seven machine tool companies here. Part of this may be ascribed to limited resources as compared with the large multinational corporations, but at least as important, as will be argued in more detail further on, is that the nature of these distinctive, specialist firms means that they have built-in choices for maneuver. Globalization goes both ways, and if a company is possessed of a unique combination of skills, capabilities and techniques, the world may well be induced to seek it out. Hence, according to its website and other data, Hamai has not so far invested in manufacturing or sales bases overseas, yet already in 1980 it contracted to produce original equipment for Pratt & Whitney, the American aircraft engine maker (www.hamai.com).

Box 7.1   Koike Sanso the industrialization LME

In the terms of our typology, Koike Sanso is an 'industrialization LME'. Machine tools are inextricably enmeshed in the industrialization process and, being founded in 1918, Koike came into existence at a time which qualifies it for such a designation, as well as plumb in the middle of the first broad phase of Japan's industrialization extending from the beginning of Meiji in 1868 to 1945. Over a

---

Box 7.1   Koike Sanso the industrialization LME – *continued*

period spanning virtually 90 years since then the company has concentrated its efforts within a tightly defined range for which it has been rewarded with an unquestionable leading status within its chosen area of core competence in Japan and, beyond its national borders, a progressively extending prominence worldwide. In its approach and what it has become, therefore, Koike neatly embodies the universal LME, the Japanese LME, and the globalizing LME.

As we have seen, the universal LME is a dyed-in-the-wool specialist. Koike began life as a firm to manufacture and sell gas, welding and cutting equipment as well as to purchase and resell oxygen gas, carbide and the like. In essence, nothing has changed, except needless to say continuous product upgrading, elaboration and development, and an enhanced awareness of the environment. There are now four broad divisions of activity: cutting and welding equipment, gases and gas-related equipment, environmental systems, and welding material and merchandise supplies. With respect to the first of these divisions, which can be regarded as the company's inner core, it now holds the top position in Japan for gas-powered machine tools, which include welder-cutters and plasma laser cutters. Moreover, it claims to provide the world's widest selection of such products, ranging from hand-held torches to portable and CNC machines. These CNC machines also evidence Koike's advance into supporting software with its automatic programming systems to prepare NC data and CAD systems as a *de rigueur* exercise for any LME in keeping in step with the times.

As for the Japanese LME, attention is to be drawn here to the way the company opts to organize and extend its business. The first thing to note is that Koike Sanso Kogyo*, to give it its full official 'English' title, is by no means large; it employs just a little more than 300 people. Rather than economies of scale, therefore, the company's primary means of growth is through economies of scope, that is through coordinating activities with consolidated subsidiaries and affiliated companies. Representing the former are Gunma Koike, which manufactures for the parent company gas cutting equipment, and Koike Medical, engaged in the production of gases for medical treatment. Circling them are affiliates with varying degrees of association, comprising some 30 in all. Prominent among them are Koike Tech, which doubles for Koike in producing gas, plasma and laser machines, as well as its welding equipment, while also being responsible for the installation, maintenance and servicing of the

Box 7.1   Koike Sanso the industrialization LME – *continued*

group's products in Japan. Most of the rest of the affiliate companies flesh out the marketing and sales operations, notable among which is Sangyo Sizaisyoji, which handles Koike's full range, including machines, welding equipment and gases, although others like Kawasaki Oxygen and Chugoku Acetylene are obviously more specialist. Then, constituting another ring beyond the subsidiaries and affiliates, are the 50 Japanese manufacturers of welding-related supplies with which Koike has concluded tie-ups to ensure that it can satisfy every conceivable need in the welding sector.

The Arcade Herald can introduce us to Koike, the globalizing LME. Arcade in the state of New York is the home of Koike Aronson, the group's American consolidated subsidiary established in 1985, and the Herald's edition of August 10, 2006 carries a photograph of a young turning-teenager fourth-generation Koike, Yannick, posing with Jerry Leary, the President and CEO, as they commemorate the beginning of the expansion of the Arcade facility. Leary is not a Japanese name, but then neither is Yannick; this family LME, with chairman Yasuo and president Tetsuo, seems to be intent on becoming international from the top down and from the inside out. More seriously, as we have seen, Koike's foreign direct investment strategy – entailing as it now does Koike Aronson in the United States, Koike Korea Engineering in South Korea and Koike Engineering (Tangshan) in China, all for manufacturing and sales, and Koike Europe specifically for sales – typifies the prudent, premeditated approach of not a few universal and Japanese LMEs. But like many of them also, Koike has another very important string to its globalizing bow: distributors. Koike has some 80 of them worldwide. Looking at the world location map, from Arcade they hop, step and jump across North America, then cascade down through Central America to key coastal cities in the southern continent. From Japan they curve through the Korean peninsular into Beijing, Shanghai and Guangzhou down through Malaysia via Singapore to Indonesia and then take the leap to Australia and New Zealand. They are likewise to be found in India and Saudi Arabia. As one would expect, they are clustered quite densely across Europe and into Russia, but they are also present in Egypt, Morocco, Kenya, South Africa and other countries in that continent. Thus does a true LME globalize.

* This can be translated as Koike Oxygen Industry.

www.koikeox.cop.jp

Summary

During the 1930s, Japan became increasingly preoccupied with military buildup. It was by then both a modern industrialized state in Western lines as well as being path-dependent based on its particular course of development. With the buildup, the government through the newly established Ministry of Commerce and Industry sought to control and channel production. Intended outcomes resulting from this were augmented ministerial influence and corporate managerial predominance. Arguably unintended was the more important production role assumed by small companies, some of which would become LMEs in the high-growth era of the 1960s.

However, all this did not go uncontested and central planning did not necessarily achieve what it set out to do. The conflicting interests of the military, the bureaucracy, big business, and small firms saw to that and compromise was the result. Moreover, it was demand more than anything that kept the small firm afloat and active, and indeed new entrants proliferated. This was particularly so for machinery, including machine tools, which descended into areas of minute specialization as per the *byora* (fastener) industry. Given the volatile conditions the large assemblers favored outsourcing which induced interconnections between different sectors of the economy and constrained in-house diversification while encouraging the concentration of specialization. One example was the emerging automobile industry which from the outset relied heavily on subcontractors, some closely tied to the assemblers and others with considerable independence.

The machine tool industry can be regarded as an LME industry in Japan because a significant number of the companies involved are medium-sized and specialized. It is at the heart of *monozukuri* and has been pivotal to the economy since the early 20th century. The LMEs which have emerged in the industry owe much to the smaller firms supplying them. The seven firms presented here have achieved that status, having been established in the prewar period. They have remained compact and unswerving from their respective core competences. Koike illustrates this, as well as the internal growth and expansion involved to enable it to achieve this end.

8
Growth and the 'Dual Economy'

Postwar determinants

The postwar years to the end of the 1980s, but particularly what can be regarded as the golden era during the high-growth phase from the end of the 1950s to the beginning of the 1970s, were very good years for the gestation of leading medium-sized enterprises in Japan. In the process there were three determinants specific to the period which assisted the outcome. The first of these was the emphasis placed on economic growth to the virtual exclusion, in the early stages at least, of broader social considerations. In this environment, the hopes of equitable dealings between all actors in industry and commerce were unceremoniously quashed in favor of an unremitting drive to catch up with and achieve parity with the most advanced economies. The second determinant was policy, and specifically industrial policy and small business policy, which over time merged to become in practice virtually indistinguishable. During the initial stage, first industrial policy and subsequently and briefly small business policy as an adjunct, were positive boosters of growth and therefore of the type of company that could make this happen, to wit the LME in many cases. Third was international pressure consequent not least on the first, for to gain the higher reaches of credibility Japan had to prove its standing by becoming an accredited and ranking member of the world's new postwar economic mediators – GATT (the forerunner of the World Trade Organization), the World Bank, the IMF, and the OECD. For gaining such status it was incumbent upon Japan to conform to the demands of these organizations, most notably trade and financial liberalization, which conversely menaced growth ambitions predicated on protectionist inclinations.

The primacy of growth

Initially, in the immediate aftermath of the Second World War, the samurai were confronted with a challenge to their authority and preferred ordering of things. Encouraged by the New Dealer spirit prevailing in the early years of the Occupation, industrial 'democratization' was all the rage. What this amounted to in the formula we have employed is that the artisan – or his political spokesman at least – saw his opportunity, during a brief euphoria of liberating ideals, to flatten the hierarchy over which the samurai presided. His small firm would have access to banking capital and materials on a par with that of the large corporation. He would practice his trade in a competitive environment far less encumbered by bureaucratic dictation and *zaibatsu* exploitation.

To the samurai, needless to say, this was tantamount to upsetting the conventional apple cart. On his side, moreover, was the simple fact that both money and materials were, especially over the first five years after Japan's capitulation, in desperately short supply, which meant that some form of selective rationing was inevitable. Hence, this excursion into unfenced territory by the Japanese was to prove as ephemeral as their much beloved cherry blossoms. As it turned out, the 'democratization' as advocated by avid reformists was simply not on the cards. A reversion to past practices was to be the order of the day. What is more, this was to be presented as being in the paramount interest of national recovery. Economic growth took absolute precedence.

Coal and steel and ... exports

Priority was initially afforded to the basic industries of coal mining, steel, electric power, shipbuilding and chemical fertilizers, to be followed progressively by petrochemicals, synthetic fibers, general machinery, plastics, automobiles, machine tools and electronics. In devising this program the samurai could be reasonably confident that it could succeed. First, Japan's own experience and experimentation with military equipment during the war years had bequeathed a legacy of capabilities which could now be applied to the development of new products, notably in heavy industry which was to be the all-important driver of growth. Second, Japan could absorb and put to use the backlog of Western technologies which had been largely denied to it since the 1930s. Moreover, there was greater flexibility involving more participants. Although the *zaibatsu* had had their trademarks returned they were not the creatures of yore. In fact, shorn of erstwhile family dominance, they were on their way to becoming *keiretsu*, relatively loose-knit groups of companies, beholden to a main bank but with considerably more autonomy, to be joined in short order by the likes of Sony and Honda. So, although remote from the ideals of the democratic purists, there was an atmosphere of greater openness and acceptance embracing a broader swathe of society within the ambit of leadership.

At the same time, the Ministry of International Trade and Industry (MITI) – successor to MCA – was treating the more immediate macroeconomic ailments it beheld with dosages of supply quotas, price controls, subsidies and financial assistance. The chief element of this planning as it evolved was the prioritizing and selecting of technologies which were then imported largely by means of licensing by designated firms from foreign suppliers, rather than letting the latter set up and directly invest in Japan. Growth was to be as much as possible internally stimulated. In all this MITI and its confrère ministries were hardly starting from zero. Immediate postwar depredations aside, Japan's production capacity was not excessively depleted. It basically retained its military supplies and other important industrial sectors, often involving some of the LMEs in the making already mentioned. Much was there to be redirected to peaceful ends. Just as important, many of the country's engineers, managers and other qualified personnel remained (Zhu, 1993:33).

There was also the fact that from the early 1950s the United States was embarking on a consumer boom on its way to becoming the great affluent society, trailing in its wake the rest of the free-world economies and providing Japan's exporting companies of all sizes with unprecedented markets in anything from shoes and toys to machines and tankers. At any event, these overlapping stimuli of central industrial guidance, inherited and imported endowments, and overseas purchasing power ensured that by the mid-1960s Japan was exporting in mounting volume automobiles, synthetic fibers, tape recorders and televisions. By 1960 its economy surpassed that of Canada, by the mid-1960s those of Britain and France, by 1968 that of West Germany (Yoshihara, 1994:21–2).

The smaller firm and growth

This high-gear expansion could have been accompanied by a radical transformation of the country's industrial structure whereby the large corporations, *keiretsu*-related or otherwise, simply absorbed the smaller fry along with their accumulated skills and production capacity. However, although overall the major corporations may have secured a larger portion of the pie for a time, essentially in its entirety the structure did not bend. Many small firms stood their ground, many new firms came into being, not a few grew and indeed flourished, and some became LMEs. A number of reasons can be proffered by way of explanation. To begin with, until the plain sailing of high-growth in the 1960s was attained, there was much volatility. The Korean War may have engendered a boom but this was followed by a slump and other lesser ups and downs during the remainder of the 1950s.

The speed at which recovery took place on the ups was to oblige large-scale assemblers to resort back to their prewar and wartime practice of subcontracting, while the boom-and-slump instability disinclined them from assuming the extra employment burden inherent in internalization (Evans, 2003). At the same time, however, it was felt imperative for them to attract and retain well-educated middle- and upper-management potential as well as talented engineers and mechanics in order to absorb, utilize and commercialize the more sophisticated emerging technologies. Corporate commitment to life-time employment was revived with renewed vigor in what amounted to translating what had been a variable cost into a fixed cost (Gerlach, 1992). So flexibility had to be sought with respect to other costs to offset this internalized fixed labor cost, and one of the solutions was to outsource the supply of materials and components to SMEs who paid their workers considerably less.

Inducement for the large assemblers to adopt such an approach was further compounded by more purely economic pressures. Foreign currency control by the Japanese government authorities did not just entail the curbing of imports of finished goods. It was also insisted that as far as possible the parts necessary for assembling finished goods in Japan be produced domestically. Rather than assuming the burden themselves, the assemblers relied on those many SMEs, notably those in machinery-related production, who had already from the prewar era made their mark in the munitions industry (Fuji Research Institute, 1998).

Then, with the establishment from around 1955 of mass production in large assembly plants in industries such as household electrical appliances and automobiles, came the call for a new type of SME capable of keeping pace and producing in volume and to precise specification. By the start of the 1960s, moreover, as the heavy and chemical industries grasped the initiative from light industries, technologies became increasingly sophisticated and specialized. Accompanying this, there was a greater need for subcontracting companies who could be trusted with calibrating their work with such technological advances and the growing complexity of the manufacturing process, from which emerged SMEs highly specialized in the production of particular parts and the execution of particular services. This ultimately led to the restructuring of subcontracting systems whereby certain companies were selected for their ability to take orders for units and subassemblies and they became top level subcontractors. Below them were second and third level subcontractors within a subcontracting structure which had become layered (Morishita, 1996). Many of these top level providers were transformed into LMEs. Moreover, in reacting with timely alacrity to the need presented – as well as taking advantage of official assistance available – the logic of subcontracting had, as it were, to a considerable extent outsmarted the logic of large-scale integration, as the artisan reasserted his right to be reckoned with.

Put more specifically, with the advent of 1960s Japan had entered its era of high growth. This brought with it market expansion both at home and abroad which encouraged the upgrading of the industrial structure, and this in its turn ushered in new industrial sectors, new products and specialization of production. On the one hand, rising technological demands required more capital, a somewhat larger scale of operations and a more complex organizational structure for those medium-sized enterprises which were to supply parts and then move on to produce specific machine tools and whole machine units. On the other hand, there was an elaboration of the division of labor which was an important contributing factor in allowing budding LMEs to become true specialists. LMEs could increasingly farm out to a growing number of diversifying SME specialists, which allowed the LMEs to concentrate more intensely on their specific core competences. Specialization flourished with growth, in other words, including among SMEs whose mounting expertise served to support the evolution of the LME. This was nowhere more so than in and around Higashi-Osaka, where the *byora*, or fastener, makers clustered. The leading industries in postwar Japan included those making bicycles, household appliances, ships, cameras, motor bikes, machinery, automobiles and construction equipment. All needed screws, bolts, nuts, and rivets.

The making of policy

The point is that, given the industrial structure as it evolved, specialization as practiced by discrete production units of varying degrees of autonomy came to comprise an essential ingredient. Small firms could become adept at it, and it was the seedbed of a growing number of LMEs. Industrial policy has its detractors, especially in the case of Japan, where arguably its supposed application only appeared to be successful because of rapid economic growth, which would have occurred anyway. However, in the broader sense policy as an expression of central government and leading business preferences did influence the shape assumed – or reassumed to a considerable extent – by Japan's postwar industrial structure. Policy commenced by rehabilitating a system dominated by large corporations. That was through to the middle of the 1950s, at which juncture it started to become much more apparent that the large corporations needed the input of parts and components from smaller firms if the national targets for developing specified industries were to be met. But this concession, such as it was, was premised on the idea that smaller firms were co-opted into policy-building to serve the cause of the large business interests. One corollary of this approach was that the more capable and stronger smaller firms were in the position to benefit from a preferential situation to hone their specializations and perchance elevate themselves to LME status.

Leveraging the LME

SME policy from the late 1940s into the 1960s is a tale of progressive backtracking from initially stated ideals combined, latterly, with a phased incorporation of small business into macro national industrial policy. However, that having been said, it was under these circumstances that administrative policy made its most significant contribution to the evolution of the LME. With the large corporation system revived and industrial democratization rumblings subdued, the administration, spearheaded by MITI, could embark on its industry promotion strategy (*sangyo ikusei seisaku*). MITI shaped the course of the country's industrial development by channeling investment funds into areas thought to have high and rapid growth potential, most notably heavy industry. Self-evidently this was an embryonic industrial policy presaged on large corporation dominance.

Yet even as it was being drafted a sense of official embarrassment was brewing as the contrast between the modernizing large corporations on the one hand and the pre-modern family-run micro businesses on the other became starker. Within the same country there appeared to coexist two vastly divergent economies, one advanced and on the move and the other backward and stagnant. The immediate reaction was to attempt to alleviate the problem, both through legislation to stamp out the worst abuses meted out by the large assemblers on their small suppliers, and in the form of financial assistance. In addition to facilitating funding, subsidies were provided through trade associations to their members for equipment modernization. Thus were the first steps taken in weaving the smaller firm into industrial policy. Note though that the key mobilizer of this conservative start on integration was *selection*. Over the decade from the mid-1950s to the mid-1960s, selection was the driving force during the initial stage of industrial policy, whereupon it was succeeded by the key concept of *modernization.*

Both of these processes were instrumental in the evolution of LMEs because they contributed to positing them within a structure still demonstrably hierarchical. In other words, the industrial structure was at that juncture far from evincing the networking and spectrum-like manifestations it was to assume by the end of the 20th century, and yet at the same time LMEs began more assertively to insinuate themselves upon that structure. One factor inducing this outcome was that in the legislation enacted firm size was still a factor, assistance being afforded to those SMEs who had already achieved a certain scale of production. That is to say, the SMEs who could benefit from policy were those who had attained a given level of operations off their own bat which in turn made them stand out in the crowd (Kurose, 1997). They were already of the type which the administration advocated: larger SMEs capable of realizing economies of scale and mass production of mechanical or electrical components, for example. It is not an exaggeration to say, therefore, that policy backed the 'winners' and encouraged them to go faster. Many of these, like machine tool maker Makino Milling, condenser maker Mitsumi, and painting machine maker Anest Iwata – all postwar newcomers – were well on their way to establishing their LME credentials.

More specifically as far as the LME was concerned, what had happened was that between the industry promotion laws of the mid-1950s and the legislation for SME in 1963 the LME, and notably the LME supplying intermediary inputs, had been enlisted as a cardinal factor in a *national* industrial structure, the building of which had been the overriding preoccupation of the policymaker. What is more, modernization was becoming the consuming ambition. Having started with large corporations, attention gravitated downwards to the smaller firms. But this effort was aimed exclusively in the first instance at specific industries. For the SME was no longer the singular entity of the original Antimonopoly Law of 1947; 15 years later industries were the key and it was what industry the SME was in that counted. The modernization of the SME was in the interest of the large assembler which relied on it for parts, moulds, stamping and so on. The SME had to achieve advanced-country standards for its parent client to succeed in the international arena. The SME Basic Law then, when it came, was to serve the purposes of a national industrial structure with its renewed oligopolizing tendencies, intent as it was on modernization at all levels necessary to assert international competitiveness.

Moreover, modernization was a fast moving target in an age where technological advances were progressing and multiplying at an unprecedented rate. Keeping up demanded intensified specialization and this is where many Japanese LMEs – long-established ones as well as new ones – made their mark as indispensable contributors to their particular industry and thus to the national cause. The LME was now *within* the definition of the national industrial structure because it was essential to the industry or industries in which it participated. As such the LME may well have taken advantage of low-interest loans from the Japan Development Bank or other semi-government financial institutions, availed itself of special tax incentives, and benefited from foreign exchange provisions for importing technology, machinery and materials, but it was nevertheless an adept within a private sector which was, by the 1960s, the main impetus for growth. Furthermore, regardless of the proportions of industrial policy, individual corporate striving, and domestic and international economic growth constituting the admixture that had brought it to the position it now was, the LME was *needed* if policy was to be seen to be succeeding. And success was essential in a world of mounting international pressures.

International pressures

The pervasive international pressure which Japan has had to confront over the past 60 years or so since the end of the Second World War has been the demand to conform to international practices as represented in the first instance by the necessity to comply with market opening stipulations as a member of the IMF, the former GATT and now the WTO, and the OECD. This has been joined since the 1990s by globalization-induced standards such as ISO and international accountancy rules, as well as alien and increasingly intrusive practices and institutions epitomized by mergers and acquisitions and investment funds. In addition, there have been three jolting experiences which have done much to reset the course of Japan's economic development in ways far from detrimental to the progress of many of the country's LMEs. The first of these was the shift to floating exchange rates in 1971 which had the immediate effect of raising the value of the yen against the dollar. Then there were the two oil shocks of the 1970s, particularly the first one of 1973 which propelled Japan on a mission to acquire energy-saving technologies. Finally came the Plaza Accord reached by the G5 Summit in 1985 for realigning the world's major currencies, which over the subsequent two years resulted in familiarizing the Japanese people and not a few non-Japanese with the term *endaka* – yen appreciation – as that currency once again soared to unprecedented heights.

Internal reactions

Looked at overall the main immediate effects of market opening preparations, or liberalization, was an accelerated concentration of capital and further confirmation of the large corporation production system. Under pressure from the imminent entry of foreign capital, there commenced in the 1960s a wave of corporate mergers among major Japanese producers – in steel between Yawata and Fuji, in automobiles between Nissan with Prince to take two of the best-known examples – which peaked as capital liberalization implementation was being finalized in 1973. Trade liberalization put to rest the government's ability to directly control import allocations, which meant less say over corporate investment, and diminished its ability to defend domestic industries against imports.

This official nervousness notwithstanding, however, and again looked at overall, liberalization actually accelerated the introduction of new technologies. Foreign firms brought in their own technologies which were either shared in joint venture arrangements or which stimulated emulation. For Japanese firms – LMEs very much included – who absorbed advanced technologies emanating from abroad the effect was indisputably positive. The enhanced foreign presence also pushed the smaller Japanese firms to improve their own production technologies, upgrade product quality and strengthen their management practices. At the same time, the arrival of foreign capital did not lead to the control of markets by outsiders. Japanese firms were afforded the leeway to upgrade without being taken over.

The internationalizing arena

International pressure was also manifested in more strictly economic terms. As the 1960s drew to a close, beyond its shores Japan was confronted with those who sought to follow in its path, notably its East

Asian neighbors South Korea, Taiwan, Singapore and Hong Kong – the four Asian NIEs – in the first instance, and those who wished to curb imports emanating from Japan, primarily the United States and increasingly Western Europe. The toys, shoes and textiles shifted in large volumes to new East Asian production sites. 'Voluntary' export restraints, already in effect for cotton textiles and crude steel, were to be progressively imposed on Japan by the United States through the 1970s and 1980s for specialty steel, color televisions, automobiles, semiconductors and machine tools.

With international competition mounting in labor-intensive, low-technology industries, in which wages were effectively falling globally while Japan's were rising, Japan's erstwhile comparative advantage was being unceremoniously undermined. Rising domestic wages and other costs, coupled with the desire to circumvent saber-rattling protectionist moves in the advanced countries, were driving brand-name Japanese manufacturers overseas. Adding value was one alternative for Japanese firms of all sizes, for which rationalization, labor-saving and automation proceeded apace. Expansive moves were apparent in metal product manufacturing, general and electrical machinery, vehicles and precision instruments, all of which required a high degree of processing and yielded high value added. Put another way, the economy was being conspicuously recalibrated along a course where the inclinations of the LME lay.

Confirmation of role *circa* 1970

The above three determinants – the primacy of growth, industrial policy, and international pressures – did not create LME specialists, but they did stimulate an eventuality. By 1970, in the main, those LMEs which arrived at that juncture from past evolution and those that were born during the first 25 postwar years were not the progeny of a top-down reordering; most of them were not mere spinoffs of large corporations – although there have been quite a few significant players formed in this way. Neither were these LMEs standout beneficiaries of policy largesse. Firms of all sizes took advantage of subsidies and other favors stipulated by legislation for promoting industry. True, those LMEs or potential LMEs engaged in machine tool and electrical machinery manufacturing did particularly well in that the legislation enabled them to bring in state-of-the-art equipment, but this was simply because they happened to be involved in these industries, not because they were earmarked on account of being or promising to become LMEs. So quite the contrary of being seen as acquiescent accomplices to state planning, LMEs are far more appropriately regarded as stalwart representatives of the private sector acting relatively freely and essentially independently of the administration in a competitive arena functioning within the parameters accepted and adhered to by business practice in Japan. Their initial essays at internationalization and consolidation on their home turf around that time can offer some insights into the environment in which they were then functioning.

Going international

Overall through the 1960s most Japanese firms looking at the outside world tended to see two options in overseas market potential. First was the immediate vicinity of noncommunist emerging economies of East and Southeast Asia, notably Taiwan and South Korea. Second were the advanced countries of the West: the United States and then Europe. At the beginning of the decade East and Southeast Asian countries were simply deemed as individual markets which were protected but growing; tapping these markets while getting round the heavy import duties meant setting up locally. By the end of the decade, among these countries and Taiwan in particular, industrial policy had done an about face, eschewing protectionism and import substitution for aggressive exports and picking winners. This presented opportunities for Japanese LMEs to export parts, machinery and materials to enable the production in these emerging economies of toys, shoes, sewing machines and household electrical appliances. Shimano's 3-speed hub, by contrast, was self-evidently made for the advanced world, while Pentel's writing instruments could appeal to both East and West. At any event, exports were the main vehicle of internationalization at that point in time. Setting up shop abroad was still rare. One exception was Nippon Paint, proving itself a practitioner of the neighboring country, proximity option by establishing affiliates in Singapore and Malaysia in 1962 and 1968 respectively, and by 1970 branches in Singapore again plus Bangkok, Jakarta and Taipei, as well as a trading office in Hong Kong. But there were not so many Japanese LME factories humming beyond Japan's shores in 1970. Things were about to change.

Box 8.1   Pentel the exporter

In 1946 Horie Yukio dissolved Horie Bunkaido, a wholesaler of Japanese traditional writing brushes and related items started by his father in 1911 and where he himself had been apprenticed as a craftsman, and established in its place the Japan Stationary

Box 8.1 Pentel the exporter – *continued*

Company which was later to be renamed Pentel. His motivation initially was the idea that what the new postwar Japan needed most was education. Education meant children so the product range commenced in the same year with crayons and paints which were avidly snapped up in a society in those hard times mercilessly deprived of novelty and diversion. The craftsman turned into a salesman, rapidly carving out market share in Japan before extending with alacrity to Southeast Asia, North America and Europe. When he retired in his 90s after 53 years at the helm he bequeathed to his successor, grandson Keima, a company with manufacturing bases other than in Japan in Taiwan, France, the United States and China and additional subsidiaries and branches for sales, marketing and distribution on every continent, dealing in a range of over 140 products. Pentel owes its success to quality, innovation, and – very much so – exports.

Exporting started in earnest in 1953. That year Horie Yukio wrapped some crayons in a *furoshiki* (Japanese kerchief) and took himself off to Okinawa – then still under American occupation – Taiwan, Bangkok and Hong Kong. At each port of call he combed the yellow pages in his hotel room for likely leads. He not only opened sales offices but also held lectures jointly with well-known local artists and conducted drawing lessons for the potential clientele, the children. The following year he dispatched his staff to the Southeast Asian locations to do likewise, pushing sales in retail outlets and department stores, teaching the children in schools and churches. The next stop was the United States, reached in 1958 having duly registered the company's first world's first in 1955: a crayon allowing for easy color blending. Sales agents were appointed in Hawaii, Los Angeles and New York and the same demonstrations were performed, notably in Hawaii where Pentel was featured on radio and television. The story also goes that, not being able to get into a convention in Los Angeles because of his limited English, Mr. Horie stood outside dishing out samples and visiting cards. A mythical touch one might say, but one backed by quality which told.

For in 1963, Pentel came out with another first, the fiber-tipped Sign Pen, a cross between a pen and a marker. This won Time magazine's Product of the Year award. It was used by President Johnson to place his signature on the State of the Union address, a moment captured

> **Box 8.1   Pentel the exporter** – *continued*
>
> on Newsweek's front cover page. In 1965, because Sign Pen did not leak under extreme atmospheric pressure changes, it was taken on board Gemini space flight numbers six and seven. But to return to the Johnson photograph. The reaction was explosive, in America, around the world, and back in Japan. To keep pace with demand two new plants were built in 1963 and 1964, then two more in 1965; in response to stardom Pentel Hong Kong came into existence in 1964, Pentel of America in Chicago in 1965, and a branch in Paris in the same year, to be followed by branches in Australia, Germany and Britain before the end of the decade. Exports, supported by innovation and quality, had internationalized a Japanese LME. As testimony to the export effort, in 1966 Horie Yukio was personally presented with the Prime Minister's Award for Exports; innovation and quality were to receive incontrovertible recognition when the company was presented with the much-coveted Deming Award in 1976. But even before that, in 1970, Pentel had achieved a form of virtual eternity. For to commemorate the Osaka Expo in that year a time capsule to be opened 5,000 years hence was prepared for which Pentel's Sign Pen was selected as the representative writing instrument of our era. It is not for us to surmise whether Pentel will still be exporting 5,000 years from now, but what it *was* exporting in the second half of the 20th century will be there for distant posterity to peruse.
>
> Pentel (1996), Evans (2003)

Consolidation at home

Having been buffeted initially by erratic twists of fortune in the 1950s, Japan then benefited handsomely from affluent markets which complemented its own halcyon efforts during its high-growth period in the 1960s. Speed was the essence; adaptation to change was the key. Such an environment was fertile territory for LME incubation, as we have seen. But we have also noted that the genuine LME is a creature of survival and growth. Some of the means by which this had been accomplished by 1970 are outlined in Figure 8.1, with reference to five veterans of the *genre*. These five are cited to illustrate typical LME strategies: enlarged capitalization, plant expansion especially in the provinces, the establishment of subsidiaries in preference to internalization, diversifying the customer base, and the building of a nation-wide presence.

It is intriguing to note that heading the pack, having been selected for its aggressive buildup in capitalization during the 1950s and 1960s, happens to be the oldest company we have encountered in these pages, the perennial LME Fukuda Metal Foil & Powder. Traditional core competence is not phased by modern business practices it would seem. Progressive plant investment by Shinagawa Refractories, on the other hand, reflects the need for client proximity given the nature of the product. There again, as noted on a number of occasions so far in this book, the Japanese LME has often been prone to manifest growth not through internal expansion by initiating new divisions but by setting up wholly owned, separately incorporated affiliates. This is what Ezaki Glico did as early as 1956 when it inaugurated two such affiliates, thereby creating a comprehensive food corporate *group*.

An overemphasis on *keiretsu* dominated by giant corporations and on pyramidal subcontracting can distract attention from the variety of customer-supplier relationships in Japanese industry and indeed the degree of supplier autonomy. Imasen has always been a supplier to the automobile industry, but equally it has always been independent, a fact that it demonstrated with aplomb during the immediate postwar decades under discussion. Within a timeframe of around 13 years Imasen managed to snare contracts with a fair chunk of the domestic auto fraternity as enumerated in Figure 8.1. This illustrates what was beginning to transpire for not a few LMEs during that period. Whereas the *keiretsu* assembler could draw up a chart with it at the pinnacle of a subcontracting pyramid, suppliers like Imasen could just as easily depict a pyramid laid on its side with itself on the left from which splayed spokes describing a half moon comprising clients engaged in the same industry, as above, or increasingly in different industries: automobiles, electrical appliances, and medical equipment, for example. Hisamitsu Pharmaceuticals demonstrates that it is still possible for an LME to have a nation-wide presence while remaining based in the provinces, in Kyushu in its case.

The 1970s and the challenge of low growth

However, these were strategies adopted in the good times of high economic growth. Those days were coming to an end. The retreat from fixed exchange rates, together with hiked American import duties, essentially put pay to the remnants of 'third world' Japan: cheap exports of toys, shoes, textiles and sundries were no longer tenable. Just as lasting in its impact was the action by the Organization of Petroleum Exporting

Figure 8.1   Pointers to Survival and Growth, 1945–1969

Pointer	Company (Founding)	Product	Action
Capitalization	Fukuda (1700)	Metal foil and powder	(Having been ¥500,000 in 1936) Raised in 1953 to ¥10 million 1957 to ¥50 million 1962 to ¥120 million 1964 to ¥160 million 1967 to ¥200 million
Plant	Shinagawa (1875)	Firebricks	(Having already established plants outside Tokyo in Fukushima and Okayama (3)) In 1960s set up extra plants in Hyogo, Kanagawa, Hiroshima, and (again) Fukushima
Subsidiaries	Ezaki Glico (1919)	Confectionary	In 1956 established 2 affiliates: Glico Dairy Products Co., Ltd. Gilco Foods Co., Ltd.
Customers	Imasen (1932)	Automotive parts	Concluded supplier contracts in 1949 with Toyo Kogyo (Mazda) 1950 with forerunner of Mitsubishi Motors 1957 with Honda (motorbikes) with Yamaha 1960 with Fuji Heavy (Subaru) 1961 with Hino Motors (trucks) 1962 with Nissan
Going National	Hisamitsu (1847)	Transdermal patches	Headquartered in Kyushu, set up representative offices in 1952 in Osaka 1957 in Tokyo 1966 in Nagoya Stock exchange listings in 1962 on Tokyo Second Section and Fukuoka 1964 on Osaka Second Section

Source: Websites of the companies in question.

Countries (OPEC), starting in October 1973, to treble the price of oil. Having relied on oil for 90 percent of its energy requirements, almost all of which it imported, Japan was caught totally unawares. In the following year its GNP showed a negative growth rate, for the first time since the war, of 1.3 percent. However, it bounced back to 5 percent by 1976, remaining around that figure for the next two years, and although the second oil shock late in the decade had some dampening effect on the economy, government and industry were much better prepared (Yoshihara, 1994:22).

The advent of the 'new' LME

There also commenced a disentanglement of the erstwhile relationship between government and industry. In the recounting so far, it has been possible to link the evolution of the LME, albeit partially, to administrative planning and guidance. Such governmental intervention and involvement had come under mounting criticism, though, not least from an increasingly autonomously mobilizing private sector. The upshot was that MITI rebounded from the oil shocks decked out in a new garb; the nitpicking planner had reemerged as the zealous practitioner of forecasting, reorienting its activities to the processing and diffusion of commercial and industrial information (Nonjon, 1999). The LMEs which came into existence in the 1970s and 1980s, therefore, were even less the offspring of preconceived need in a bureaucratically defined catch-up mode. They were far more the result of exigencies perceived in industry itself, and particularly those industries approaching or bestriding the technological frontiers.

The situation called for a new type of LME. But when we say 'new type' here, although it includes companies created from the 1970s on with a vision fashioned by the evolving demands of the time, it also encompasses the many already extant LMEs who successfully adjusted to the new environment premised on knowledge and flexibility. After all, all the LMEs we have encountered so far, some truly venerable, are still alive and kicking to this day. What had happened was that the primacy of mass production and mass marketing as the means of growth for the LME was being joined, if not replaced, by demands for diversification and individualization. While scale was the crucial ingredient in the 1960s, by the 1970s, and even more so in the 1980s, what successful LMEs had in common was diversity of production, albeit within a carefully delineated chamber. This was taking place under circumstances where networking was assuming a conspicuous reality. The more effective smaller firms were relying less and less on specific clients and broadening their customer base.

LMEs, as industrial suppliers as well as end-user manufacturers, were at the apex of this: specialists, but not simply mass producers. They now occupied this superior foothold because, among other factors, they had strengthened their R&D and design capacities, acquired a more astute sense regarding market trends which told them what to target and what to discard, professionalized their management, made constructive use of the external economy including available SME expertise, and started venturing along the internationalization trail. Moreover, they gained in independence. Although many of them had made impressive strides during the 1960s, they nevertheless played a supplementary role within the large corporation industrial structure. But the transformation to the knowledge-intensive paradigm of the 1970s inevitably led to a greater store of proprietary intellectual assets. Availing themselves of this, LMEs started to propel themselves beyond being simply *needed* in an industrial structure essentially not of their own making to become significant contributors to a new order.

The chamber of our stylized LME was now more demonstrably established than ever before. The status of such LMEs *vis-à-vis* large corporations was far more assured; many of them operated in totally different markets while just as many effectively competed with their larger counterparts by way of their own technologies and sales routes. Confident in the innate viability of its core competence, this breed of company could afford to diversify based on accumulated technology and committed human resources. It could perhaps enter new domains with its existing skills. In doing this, it increasingly had at its disposal computer aided design and manufacturing, factory automation and other currently progressing tools. In addition, just because of its stable home-grown base, it could afford to – indeed had to – pursue linkages going beyond existing frameworks to embrace inter-industry opportunities, as well as to enter into joint research with large corporations and university institutions and to dexterously exploit the advantages of outsourcing. The emphasis was not so much on economies of scale, but rather economies of scope.

Box 8.2   Ishida – 'new' from Kyoto

Ishida, based in the former imperial capital city of Kyoto, is nowadays a prominent manufacturer of weighing and packaging equipment, with which it serves the manufacturing, inspection, distribution, and retail industries. For such equipment it boasts 70 percent and 80 percent world and domestic market shares respectively. Its early existence followed a by now familiar course. Having been founded in 1893 by Ishida Otokichi II to manufacture scales as authorized under Japan's first weights and measures licensing system, Ishida Scales Mfg. Co. Ltd. was subsequently incorporated in 1948 and the company is still presided over by an Ishida – Ryuichi – to this day. Ishida can be described as an independent family-based LME, and it remains unlisted on the stock exchange. The firm's march through industrialization has been punctuated by significant advances which have stamped its presence on the scene: in 1933, Ishida came out with an innovative scale capable of adjusting to changes in temperature; in 1959, it completed

> **Box 8.2   Ishida – 'new' from Kyoto** – *continued*
>
> the development of an automatic weighing machine; and in 1969, it brought out its first electronic pricing scale. By way of further consolidating its position at home, moreover, starting in 1958 Ishida built up a nationwide direct sales network.
>
> But what chiefly concerns us here is the 'new' Ishida that emerged out of the knowledge-intensive 1970s, extending into the globalizing 1980s and beyond. The key words denoting what transpired are modernization, internationalization and reputation. Modernization is symbolized by the incorporation of computer technology, and this was heralded with the unveiling of the world's first multihead weigher, or computer combination weigher. This landmark product, in turn, epitomized the essentials required of the Japanese LME if it was to progress in the unfolding era: combining, upgrading, diversifying. To begin with, it combined state-of-the-art technology with the attention to minutiae – the essence of *monozukuri* – where the Japanese are so strong. For this weigher initially came out of a request to automate the weighing and packing of green peppers, for which Ishida developed a machine to produce bags of green peppers of a maximum weight of 150 grams and a margin of error of two grams. Then came the upgrading for greater speed and precision so that, as against 50 to 60 weighings per minute with the original design, this had been increased to 160–180 by 1980. And as a consequence of such upgrading, the machine ventured into other realms, those of agricultural products, meat, confectionary and foods in all states of preparation, metal and rubber parts, and so on.
>
> What is more, the multihead weigher was already finding markets abroad, first in Australia in 1978. Internationalization then assumed serious proportions attendant upon marketing tie-up contracts signed with Heat and Control, Inc. of the United States in 1980 and the company's first overseas subsidiary, Ishida Europe Ltd., set up in the United Kingdom in 1985. Others were to follow, including manufacturing affiliates established in South Korea in 1988 and in Shanghai in 2000. Enhanced reputation was the inevitable accompaniment to this expansion and success. In 1988, president Ishida Ryuichi was the first non-American recipient of the 'Circle of Honor Award' presented by that country's Snack Food Association. In 2006, the Ishida's X-Ray inspection system for detecting contaminants, and further witness to the company's diversification on a theme, was adjudged the best of such products in an independent

---

Box 8.2   Ishida – 'new' from Kyoto – *continued*

appraisal by The Danish Meat Institute. During the same year, the president was interviewed by the Japan Broadcasting Corporation, or NHK, on a prime-time news program devoted to blue-chip companies in Japan. Such companies, by implication, are not all behemoths. They can be LMEs.

www.ishidajapan.com; Nikkan Kogyo (2003)

Summary

There were three specifically postwar determinants of Japan's development. The first was the overriding emphasis on economic growth. Having quashed the brief attempt at industrial 'democratization', the samurai reasserted their authority and gave economic growth absolute precedence. The economic ministries, notably MITI, acquired unprecedented authority, affording them the leeway and the power to instigate planning which entailed the prioritizing of industries and the selection of technologies for designated companies to import. However, small firms were not absorbed for the following reasons: economic volatility, the growing need of large assemblers to rely on outsourcing, and government insistence on the domestic production of parts wherever possible. This was the seedbed for increased specialization and the creation of LMEs as technologies and processes became increasingly sophisticated going into the 1960s.

The second determinant was the formulation of industrial policy which in a sense started with the rehabilitation of the large corporation system and the phased emasculation of the Antimonopoly Law. On the other hand, there was a growing realization both that large corporations needed their inputs and that the parlous state of many SMEs should be rectified. This induced funding programs and subsidies for exporters based on selection. Furthermore, some smaller firms were favored by dint of the fact that they were successfully engaged in the prioritized machinery and electronics industries. Not a few of these became LMEs in the process. Such LMEs became a recognized part of the modernizing *national* industrial structure.

The third and final determinant was international pressure. This has come from the international organizations Japan has needed to join, notably the IMF, GATT, and OECD. LMEs, however, have benefited from market opening and closer familiarization with foreign technologies as they matured and distanced themselves from bureaucratic attentions. The rigors have proved to be a testing ground for eventual internationalization. They progressed with rationalization, labor-saving, and automation to address the restructured order of what was becoming an international division of labor. As the large assemblers moved abroad LMEs followed them to cater to their immediate needs while also finding new customers among the NIEs.

By the advent of the 1970s LMEs were a constituent part of the national economy. Furthermore, some of them like Pentel were already adept practitioners on the international scene, chiefly through exporting. At home they had consolidated their position by raising their capitalization, expanding their client base, extending the plant locations, and so forth. Then, as the 1970s unfolded, a new type of LME, capable of addressing the electronic and software revolutions was called for; LMEs like Ishida.

9
Contrasting Profiles of the Modern Japanese Specialist

Diverging responses to change

Overall the Japanese LME is still characterized by four distinctive features. First, it spans the ownership-management divide. The family LME evidences the simple fact that firms are created in the main by individuals. However, as many LMEs have matured the professional manager of the business text books has established his turf, although here again, as in Germany and France for instance, that professional manager can still be family. At any event, the Japanese LME of current times effect a linkage between personal entrepreneurialism and externally trained professionalism, as well as between the large corporation and the mass of small enterprises. Second, in terms of size measured in number of employees, having achieved LME status the *core company within Japan* has remained remarkably stable. There have been exceptions, such as Sony and Kyocera waxing mighty during the 1960s and 1970s respectively, as well as, conversely, some LMEs making drastic personnel cuts during the recession of the 1990s. But most LMEs are apparently locked into a set status, partially of their own volition and partially in adherence to the conservative ordering of the Japanese business structure. As we have seen, one means of unobtrusive expansion is group networking; the other – increasingly – is foreign direct investment.

The third feature is the LME's manner of dealing with business collaborating groups referred to as *keiretsu*. Parenthetically it should be pointed out that employing the concept of *keiretsu* to describe the workings of the Japanese manufacturing system has always been flawed, not least because it implies a pyramidal rigidity which the *keiretsu* rarely achieve. That having been said, the point about the LME is that, being by definition autonomous, it is in Japanese terms a free agent not belonging to any group other than its own. It is often a supplier, but usually to a number of *keiretsu* or groups, and not overly beholden to one or two. Finally, it has been an essential element in the national 'full set'. The term 'full-set' has been used in two ways in describing the Japanese economic structure. At the micro level it refers to the individual horizontal *keiretsu* embracing all the major manufacturing and service industries. At the macro level it refers to the aim of the government administrator to have produced domestically everything that it is deemed an advanced economy should produce (Seki, 1994). Japanese LMEs, in steering clear of the former have furthered the latter: S&B Foods' curry, General's carbon paper and Tanaka Kikinzoku's bonding wire have all been elements of the national full set.

A diversifying scenario

The world of manufacturing and industry in general was becoming more complex and more heterogeneous from the 1970s on, however. New knowledge-intensive industries increasingly dependent on advanced software and information technologies were being fostered. The LME both contributed to and benefited from this complexity and heterogeneity, provided it realigned its priorities to accommodate the new rules of the game. Whereas, as we have seen, the up-and-coming LME of the 1950s and 1960s proved its pedigree by achieving a capacity for large-scale production which could not be matched by the ordinary small firm, its survival now increasingly turned on the additional skill of handling diversification and customization. This depended on the ability to master the new tools of the trade, such as value-added networks, computer-aided design and manufacture, and factory automation. Making contact, adapting to specific demand, and maneuvering fast to realize the outcome required information, so that the range of information became every bit as crucial as the range of product.

This is what began to mark the LME off from the rest of the crowd as the 1970s progressed, joined however by one more indispensable ingredient: more and more the LME was not only responding to demand but also creating demand through the development and elaboration of its own products. In so doing it was shedding its supplementary role. Instead, it was establishing itself as a sophisticated specialist, a leader in its elected field with an array of proprietary knowledge and capabilities unparalleled by those it served, while it both cooperated with and competed against the big boys. Moreover, the character of the founder was changing. A society where the nail which protruded traditionally had to be hammered down now increasingly required mavericks to head teams of mavericks, insisting on solidarity to be sure but simultaneously championing innovation. Here was the impatient, ambitious artisan denying the conservative, regulating instincts of the samurai. This in turn bred novel solutions to emerging challenges, increasingly influenced by non-Japanese practices. Different solutions engender different approaches. Here contrasting profiles as illustrated by three firms are presented to give substance to the range of approaches being pursued. Nidec represents growth through absorption American-style, Takara Tomy business consolidation German-style, and Union Tool core competence application Japanese-style.

State-of-the-art entry – Nidec

With reference to the industrial categories employed by the Japan Company Handbook and similar publications, if metal products and general machinery saw a significant number of startups in the early postwar decades, by the beginning of the 1970s there had been a marked shift to electrical machinery and precision instruments. The vogue was now electronics and optics and Table 9.1 offers a selection of the kind of company established during the 1970s and 1980s which in time became LMEs. Some of these companies developed very fast, and none more so than Nidec. Most were quick off the mark with global contacts; Allied Telesis had a sales base in the United States the year it was established and now has manufacturing units in Singapore, China and Malaysia, complemented by a sales network encompassing East Asia, North America, Europe and Australasia. Simply put, Allied Telesis was 'born global', as was Nidec in its fashion.

This table, moreover, can be taken to illustrate a couple of other salient factors. The distribution of headquarter locations reflects the familiar pattern. Tokyo (plus adjacent prefectures of Chiba, Kanagawa), Osaka (Hyogo) and Nagoya (Aichi) all feature, as also do what could be referred to as the second-tier locations of Kyoto – where Nidec is based – and Nagano. But there is some further dispersion represented by the prefectures of Ishikawa and Tottori, suggesting that it is still possible for LMEs to flourish at some distance from the main industrial centers. The second factor is that concerning founding, ownership and implied control. The top individual shareholders in the far right column are all founders and presidents or chairmen. As can be seen their personal ownership of stock ranges from the significant to predominant and in quite a few cases this is supported by holdings of other family members. The implication is that the founder is not keen on muting his ownership and

Table 9.1   New LMEs Emerging from the 1970s (as of 2004)

Company	Head Office	Estab'd	Product Line	Stock Market	Top Individual Shareholder/%
Nidek	Aichi Pref	7/71	Ophthalmic equipment	Unlisted	President/19.0
Seikoh Giken	Chiba Pref	6/72	Optical communications devices	Jasdaq	President/9.7
Kodenshi	Kyoto Pref	6/73	Optical semiconductor devices	Unlisted	President/35.1
Nidec	Kyoto	7/73	HDD, CD-ROM motors	Section 1	President/9.3
Nippon Ceramic	Tottori Pref	6/75	Ceramic sensors	Section 1	President/19.7
I-O Data Device	Ishikawa Pref	1/76	PC peripheral equipment	Jasdaq	President/30.3
Zuken	Kanagawa Pref	12/76	CAD/CAM systems	Section 1	President/16.1
Nihon Trim	Osaka	6/82	Ionized alkaline water purifiers	Section 1	President/40.6
Kyoden	Nagano Pref	7/83	Printed circuit boards	Section 2	President/27.6
Allied Telesis	Tokyo	3/87	LAN-related equipment	Section 2	Chairman/47.7

influence while he remains active. To that extent bureaucratization and disinterested professionalism do not set in. More to the point, this is not the 'Japan, Inc.' of large corporative spin-offs and 'salaryman' management. Nidec's president owns just under ten percent of what is now a large outfit all told, and this helps make him one of Japan's biggest earners.

In fact, the moment, the man and the method, intermingled, encapsulate the story of the meteoric rise of Nidec. Being established in

1973, Nidec was a scion of state-of-the-art and contemporary of Cray Research, as well as being a precocious forerunner of Silicon Valley start-ups 'born global' in the 1980s. Nidec arrived on the scene when the lodestar of destiny was indicating a new departure as it pointed with increasing insistence to knowledge and information; the acme of progress was being reposited in cerebral creation in preference to the physical and mechanical. Moreover, the first oil shock that year was to afford an extra urgency which was to translate, particularly in Japan, into a preoccupation with energy saving and miniaturizing technologies over the next decade or so. Promising though hindsight retelling may imply, however, the business Nidec was launched to engage in, the manufacture of small DC motors, was at the time being shunned by existing companies due to their depressed prices. Added to which, this tiny unknown was to encounter the snobbery and indifference of the 'samurai'-dominated, large-corporation system. A product was being made for a market which refused to be there.

The man

But this did not reckon with the man. Nagamori Shigenobu, the one and only president so far, was all of 27 years old when he founded the company with a capital of 20 million yen and a factory looking out onto farmland in the Kyoto suburbs 30-odd years ago. The salaryman's life had palled very fast for an individual who had already experienced running things for himself. Besides, contrary to the wise heads in the business establishment, he was quick to see the glowing future of small DC motors, and set about the task of making his minnow the world's number one producer. With enormous pluck and energy – he works seven days a week with just time off on New Years' day morning to visit a shrine – he has achieved this. And this all started with a ploy which has been adopted by a number of other Japanese LMEs, like Horiba with its measuring instruments and Rheon with its food-processing machines for example: Nidec gambled on a 'reverse landing'. Mr. Nagamori legged it to the much more accommodating and less tradition-bound United States where the product comes first, not the history of the company. He secured orders from IBM and other luminaries of the trade which transformed Nidec's image in the eyes of those back home (Ishikawa and Nejo, 1999).

The method

This maverick approach – for a Japanese company that is – is inherent in the method. Indeed, to a fair extent it supports the argument of those who claim convergence of business practice by companies around the world, and as such offers a somewhat contrasting picture to what is being said here about the distinctive characteristics of the Japanese LME. To start with, there is Nidec's style of production. In the order of priorities, speed supersedes quality. This is best elucidated by saying that, while quality remains paramount, speed is *even more* important. Nidec recognizes that in a fast-paced industry where obsolescence is endemic, flexible and prompt response in the engineering of proto-types and speedy delivery of the product from sites located close to the customer are of the essence. This is the reason why up to 90 percent of production takes place abroad, where its foreign clients are located and its Japanese clients have moved to. Second, Nidec likes money. By con-centrating on a very limited range of products, which has nevertheless eventually stretched over a pretty extensive customer base covering the IT, audio/visual, automobile, office equipment and home appliance industries, it has built in a cost advantage over its rivals while at the same time ensuring average profits of over 20 percent and return on equity approaching ten percent, both way beyond the rates for Japan's general electrical giants (Bouissou, 2003). Furthermore, and this is the third point, Nidec likes its employees to like money, as an incentive for effort needless to say. Promotion and remuneration are based on merit; mid-career hire accounting for 40 percent of the company's workforce is *de rigueur*, introducing new blood lured by the financial rewards and profit sharing schemes. This is capped by the 'Diamond Special Prize' comprising ¥10 million in cash awarded to the individual deemed to have made the biggest contribution to the company's profits over the year just ending (Ishikawa and Nejo, 1999).

Finally, in pursuit of profit and growth, Nidec has been an aggressive practitioner of merger and acquisition. In the United States it bought the Axial Flow Fan Division of Torin Corp. in 1984, to be followed by Power General in 1991, and then on home turf it purchased the High Precision Composite Parts Division of Seagate Japan in 1992. It has been no less active in looking to fellow Japanese firms, making inroads into the automobile and home appliance markets by acquiring Tosok, Shibaura Mechatronics and Copal and so availing itself of their expertise. Its latest significant purchase, in 2007, has been Japan Servo, a subsidiary of Hitachi, Ltd. manufacturing small motors (www.nidec.co.jp). In fact there is a striking resemblance to the strategies adopted by the universal – and indeed American – LMEs turning large like Medtronic. The one important difference is that Nidec Corporation, the core company, has remained a very lean establishment of around 1,500 employees concentrated on planning and R&D, although now the consolidated workforce is over 89,000 (JCH, 2007). Nevertheless, Nidec, the new type of LME of the 1970s, presents the question mark. Is it the prototype of the emerging Japanese specialist as it outgrows its LME credentials?

The working metabolism – Takara Tomy

When, at the end of the 1980s, Nakamura Hideichiro was revising his ideas concerning the LME phenomenon within the Japanese economy initially penned in the 1960s, that economy was apparently booming, although in fact the bubble was about to burst. Many LMEs were doing well, which to Nakamura further justified the argument for singling them out not only as distinct but also as worthy of emulation. Having said that, Nakamura was under no illusions about the invincibility of the LME; it was a status which had to be earned and then constantly tended thereafter. Failure to appreciate the importance of the knowledge-intensive shift and information technology as it evolved through the 1970s and 1980s, for example, could spell disaster (Nakamura, 1990). But there was, nevertheless, still at that time a discernible pattern of growth common to LMEs which kept on course: restrained but steady expansion of the core company, progressively accompanied by the elaboration of networks of closely aligned subsidiaries and collaborating partners, albeit driven by differing underlying strategies. However, as the economy has moved out of the prolonged recession lasting well over a decade and is more than ever integrated into the global system, the responses of the Japanese firm, including the LME, have become increasingly diversified.

Metabolism adjustments

More and more it is the reaction of the firm to its individual circumstances in a global environment that count rather than knee-jerk conformity to the assumed trends of a national industry. It is no longer a matter of filling the gaps for an economy in catch-up mode, or even inventing specific needs for a single national economy, highly significant though the world's second-largest economy may be. Rather, it is a matter of looking out into a world where the horizons are somewhat blurred and trying to figure out where one best fits at any given time. And as recovery in Japan has taken hold during this first decade of the new millennium, there has been enhanced activity among LMEs as they go about reordering what is often referred to as their '*shinchin-taisha*': their metabolism. Table 9.2 presents examples. To be stressed is that these are maneuvers in

Table 9.2   LME Metabolism Adjustments

Action	Company	Year	Details
Listing	Nihon Trim	2000	Listing on Jasdaq
		2003	Listing on TSE 2nd Section
		2004	Listing on TSE 1st Section
Merger (intrafirm)	Koransha	2007	Merger of Koransha with Koransha Trading and 4 other Koransha companies
(interfirm)	**Tomy/ Takara**	2006	Merger of Tomy and Takara to form the new Tomy
New subsidiary/ affiliate	Zuken	2000	Establishment of Zuken Tecnomatix jointly with Technomatics Technologies
		2001	Establishment of Chip One Stop jointly with Nissho Iwai and others Establishment of Zuken NetWave
		2006	Establishment of Inventure
Holding company	Danto	2006	Trade name changed to Danto Holdings; Danto newly established as a company within Danto Holdings group
Spin-off	Akebono	2000	Akebono Brake Proving Grounds spun off as Akebono Tec Corporation; ditto
		2001	Miharu plant as Akebono Brake Miharu Manufacturing Co., Ltd., Fukushima plant as Akebono Brake Fukushima Manufacturing Co. Ltd., Human resources and administrative functions as Akebono Management Service Co., Ltd.

NB: TSE = Tokyo Stock Exchange

line with perceived organizational necessity; in no case do they represent departure from the manufacturing core competence of the firm in question.

The point to be borne in mind is that growing complexity and heterogeneity have brought in their wake a diversity of organizational responses, just as with manufacturing. Flushed with success come the new millennium, Nihon Trim has rapidly scaled the stock market status ladder. The much older Koransha, by contrast, still after more than 300 years apparently feels no necessity to get listed either for reasons of recognition or financing. But it has seen fit to rein in key subsidiaries under a single roof. This tightening of association, or substantiation of the chamber, is likewise evident in various ways among some of the other examples here. Toy makers Tomy and Takara have joined forces to form a more effective unit for, in the first instance, tackling a shrinking but still highly competitive domestic market; Nidec has gained control over subsidiaries deemed essential for its drive to become a large corporation; and Danto, the tile manufacturer, has taken advantage of a revision to the Anti-Monopoly Law in 1997 which lifted the ban on the establishment and operation of holding companies, thereby consolidating its activities under a single corporate entity. There again, growth of a more 'traditional' type for LMEs extending their reach can be seen in Zuken's buildup of a network of affiliates. This latter's more centrifugal approach is then given further definitive expression by Akebono which regards the spin-off of two of its manufacturing plants as a means to enhance competitiveness, and that of its human resources and administrative functions to be in the interests of promoting greater efficiency in its non-manufacturing operations. In other words Akebono has gone about loosening, as opposed to tightening, central control.

To be sure these solutions are perfectly familiar on the modern business scene anywhere. However, the Japanese LME can be singled out so far by its will to survive intact, and this influences the choices it is prepared to contemplate. In contrasting national economic setups, the Japanese LME can most easily be likened to its German counterparts, many of which are likewise specialists and frontrunners in their self-prescribed areas of expertise. A relatively minor difference between them is that as they have braced for expanding national and then international challenges, the Japanese have tended to opt for surrounding themselves with cooperating confrères, while the Germans have looked more to merger and acquisition. But, as has been noted, non-traditional practices are intruding. So although the merger of toy makers Tomy and Takara is a modern story for Japan, it represents an instance of intra-industry realignment, common as we have seen for some decades in Germany. Regarded by the Japanese as far more preferable to takeover – especially hostile takeover, in fact the merging of former rivals in the same industry has become somewhat fashionable in Japan over the past decade as the outer world impinges and uniting is deemed the key to survival as members of corporate Japan. The toy industry is in the process of such realignment at the moment and similar attempts have been evident in the pharmaceutical, paper manufacturing, and optical instruments industries, for example. The consequences of this toy-maker merger are also modern in the global context, moreover, as further ongoing adjustment solutions are now being sought through inter-industry synergies and an augmented internationalization strategy.

The problems

The main problem the toy industry in Japan has had to confront over the last couple of decades or so is the unremitting decline in the birthrate. While the cheap manufacturer catering to the American market had been scuppered by a drastic hike in import duties dictated by the so-called Nixon Shock in 1971, the better toy makers had been able to keep going because of rising affluence, superior quality, and a native instinct for the idiosyncratic tastes of the Japanese child. But this, given the shrinking customer base, also depended on constant innovation at a frantic tempo. And this is where Takara ran out of steam in 2004. Added to which it had spread its talent too thinly, dispatching the faithful to manage new acquisitions, resulting in inadequate response at the core of the chamber when the crisis broke out. There were calls for the resignation of the president, and son of the founder, Satoh Keita, notably from the leading shareholder, Konami, a major entertainment company. According to Mr. Satoh, Konami had previously attempted to take over Takara, initially proposing a 51 percent stake, which was finally beaten down to 22 percent. However, by this time – by 2005 that is – he was already well advanced in negotiations with another company which in his view had much more respect for Takara's corporate identity and culture (Satoh, 2005:113–16).

The adjustments

That company was Tomy, another family LME currently presided over by Tomiyama Kantaro, and best known for its Pokemon character toys. Tomy and Takara merged effective 1st March 2006. Through this move the new company, Takara Tomy*, was elevated to the number four spot among the nation's toy and game-software firms in terms of consolidated sales (Nikkei, 2005a). Both companies had, moreover, continued being active individually in extending their reach, Tomy acquiring Wako, a baby-clothing specialist, and Takara acquiring Tatsu No Ko Productions, both in 2005. And here with the latter is where the synergies start to emerge. To begin with, prior to the merger, Konami's

---

*The registered English name of the merged company is Tomy, while the registered Japanese name is Takara Tomy. The Japanese name is used here to avoid confusion between the old Tomy and the new one.

22 percent share in Takara had been taken over by Index Holdings, a contents distributor. With the merger, Index became the leading shareholder of Takara Tomy and subsequently acquired a 14.5 percent stake in Tatsu No Ko. What is more, Index had also been busy in the market over recent years, making subsidiaries through share acquisitions of the celebrated animation studio, Madhouse, in 2004, and of Nikkatsu, one of Japan's oldest film producers, in 2005. The stage was thus set by June 2006 for the announcement of a synergy whereby Index and Takara Tomy intended to produce television programs and movies, with Madhouse responsible for the animation and Nikkatsu the filming, using well-known characters, the copyrights for which are owned by Tatsu No Ko (Nikkei, 2006a).

But this is not enough. Demand for traditional toys in Japan is being progressively eroded not only by the low birthrate but also due to the popularity of video games. Part of the solution has to be found in an enhanced global presence. So on 7th June 2007 Takara Tomy and the Texas Pacific Group (TPG) of the United States announced an agreement in which TPG would purchase ¥7 billion in convertible bonds to become the second-largest shareholder, with a stake of just over 14 percent and three outside directors appointed to the board (Nikkei, 2007b). This is the first instance of an investment led by a private equity group in Japan. The attitude is mutually positive with Takara Tomy's president, Mr. Tomiyama, emphasizing the need for TPG's knowhow for the restructuring, which TPG is initiating by spearheading a project committee to investigate all aspects, including purchasing, brand value, and supply-chain management. The head of TPG in Japan is quoted as saying he would like to see Takara Tomy's overseas sales rate rise to 50 percent from the current 15 percent, targeting fast-growing China especially. In other words he wants to make of Takara Tomy a truly global player. What is interesting here is that such a constructive approach towards synergizing Japanese production and American – read global – marketing is being taken by this Japanese LME.

The stylized LME – Union Tool

So the world is changing at an increasingly rapid pace and the LME must respond purposefully and speedily. In fact, an essential ingredient of its survival kit is its dynamism and its adroitness in interpreting the unfolding situation in the light of its own circumstances. How each LME responds is becoming more and more an adjunct to its own individual interpretation of events. In thus responding, there is one factor that no Japanese LME can ignore: the necessity to address globalization, both to exploit its advantages in the international arena and to defend the firm's proprietary technological and marketing turf. This could suggest an argument for convergence in which the Japanese LME resorts increasingly to prepackaged global solutions, which is certainly happening to an extent as the above two examples have imparted. But nationally engendered inclinations still run high. One way to gauge this is to analyze the role of *monozukuri* in the Japanese national persona and this is to be discussed in the next chapter. Before that, though, as the implication all along in this book has been that there is a neat fit between the stylized LME and the Japanese mode of operation in manufacturing, it is perhaps worthwhile testing its credibility. How Japanese can the company be and yet still globalize successfully? Obviously a stylization is an idealized curve while the real world obliges us to plot the actual companies in its general direction, some dispersed quite a way off from the curve's trajectory. Nevertheless, there are instances where the idealized and the reality are remarkably close. Union Tool is a case in point. Here it is superimposed on the stylized LME outlined in Chapter 4 to see how well they match up.

Entity – size, stage and type

Union Tool is a manufacturer of tungsten carbide drills used for piercing holes in printed circuit boards. It accounts for around 40 percent of the Japanese market for these drills and 30 percent of the world market. For the fiscal year ending 30th November 2006, Union Tool recorded net sales of ¥28,655 million and an operating income of ¥8,238 million (Union Tool, 2007a). This kind of performance is achieved with a workforce within Japan as of 31st January, 2007 of 703 (Union Tool, 2007b) and 1,439 worldwide (JCH, 2007). Symbolic of the company's progression to LME status was the achievement of mass production of drills in 1970. In the 1960s when it was a small company with 20 to 30 employees, Union Tool was not able to comply with an approach by IBM precisely because it could not mass produce. Now it could.

A further step was the launching of its own brand in 1976. To that point in time it had been making drills under Mitsubishi Metals' brand name, so producing under its own name was an act of *datsu keiretsu* (breaking away from large corporate group dependence) in a bid for effective autonomy (Nikkei Sangyo Shinbun, 1982a). Then there was the spreading of the load of capability and responsibility. Union Tool owed its origins and mounting success to the genius and tenacity of a single individual, founder Katayama Ichiro. However, from early on he was aware of this as an ultimate potential source of weakness (Nikkei Sangyo Shinbun, 1982b) and so set about recruiting and training quality engineering and marketing capacity, an endeavor crowned by the establishment of the company's Mishima Research Center in Shizuoka Prefecture in 1996. Added to which, the company's elevation of status presaged on rapid growth was confirmed, first by its listing on the over-the-counter market of the Japan Securities Dealers' Association, followed by listing on Tokyo Stock Exchange's second section in 1995 and its first section in 1998.

In fact, it was from the advent of the 1990s that Union Tool's fortunes reached full fruition with the heady proliferation in demand for PCBs. Here we see the stage within a stage: the now established LME Union Tool poised, because of its already accumulated expertise, to make a killing in the helter-skelter advance of the PCB business. For the PCB also represents the modern dichotomy of heterogeneity and in-depth diversification. Come the 1990s PCBs were in great demand for personal computers, mobile phones, game machines and so forth, and an indispensable tool as the miniaturization of these high-density boards proceeded apace was the ultra-hard drill capable of piercing holes of some 0.4 millimeter diameter. Moreover, in achieving its leadership in the production, upgrading and elaboration of this highly specific item, Union Tool also displayed a rare talent for artisanship in that, in addition to the product itself, all the key manufacturing and testing machinery and equipment have since the outset been designed and built in-house. Out of all this has come a global LME with eight wholly owned overseas subsidiaries, including those for manufacturing in the United States, Taiwan, and China, to complement the hub of three adjoining plants in Japan's Niigata Prefecture.

Evolution – background and process

The company commenced its existence in 1960 as Union Tool Chemical Research Co., Ltd. to develop and manufacture carbide burrs for dental applications. It comprised in reality a small urban factory (*machi koba*) run by a highly gifted engineer, but heavily reliant on Mitsubishi Metals for both materials input and product marketing. For the first ten years it could muster annual sales of only five to six million yen in the small dental equipment market. But hardship generated innovation. To develop the dental burr Mr. Katayama had to devise his own machine tools because nothing suitable was available from elsewhere. Thus was born a business philosophy which led first to independence and eventually to very high returns. The fact that Mr. Katayama succeeded in producing ultra-hard cutting tools for making dental burrs was what caught the attention of IBM. The fact that, having fallen short of the mass production requirements in the first round, Mr. Katayama had already imbued a self-help frame of mind meant that to win the second round he immediately launched on a program to make good the deficiencies by creating his own machinery for mass production of drills. This – the founder's imprint – was the basis of Union Tool's artisanal *monozukuri*. And although, since then the company has remained essentially under founder family control, it has nevertheless acquired the necessary accoutrements of corporate governance by way of the second-generation president, Katayama Takao, with his strong marketing sense, grasp of strategy, and ability to reap funding from institutional investors in the global financial marketplace (Nikkei Kin'yu Shinbun, 2002).

In other words, having started with carbide burrs at the beginning of the 1960s, Union Tool had by 1970 found its true métier in the manufacture of tungsten carbide drills to which it has now been committed for over two generations. In the decade of the 1980s coming up to the big breakthrough in the 1990s annual turnover rose by a factor of ten plus from ¥700 million to ¥7.5 billion (Keieisha Kaiho, 1990). Then by the end of the millennium Union Tool was an internationally established manufacturer churning out 10 million PCB drills a month. Its commitment to quality is reflected in the 20 checkpoints for specific tolerances that each of these 10 million drills must go through before completion and shipment. Its drive for excellence in its proprietary technology is illustrated by it being the only company capable of manufacturing drills as small as 0.015 millimeter in diameter, finer than a human hair. Its insistence on superior equipment and personnel is demonstrated by the way these two are joined at the hip. To take the improvement suggestion system adopted by the company to indicate what this means, no less than 95 percent of the 50 to 100 suggestions that emanate from the workforce every month are directly related to what transpires on the shop floor, that is to say how the proprietary machinery and equipment are best employed in the processing (JMAM, 2001).

Stated more broadly, Union Tool's *monozukuri* is presaged on *hitozukuri* (making people) where understanding of what is happening in the factory is the key, and this is what the young entrants are taught for their initial three to five years. Thereafter, workers continue to be rotated through the manufacturing plant, the technology department, machine tool assembly, and maintenance to enhance companywide mutuality.

Knowledge, not physical strength, is seen as the essential requisite for the making, adjusting, maintaining, changing, and improving of super-fine, ultra-hard drills (JMAM, 2001). The employees learn to accomplish what Union Tool calls a 'triangulation of production', which connects together drills, materials and machines, not as lifetime protected drones but as highly motivated contributors to the company's success. In so doing, they are integrated into the unity of the company and its core competence, the kernel of which is what sets Union Tool apart: the ability to insert an expensive tungsten carbide drill bit, or blade, into a cheap stainless steel shank in a process called shrinkage fitting (*yakibame*), thereby reducing the use of tungsten to the minimum.

Complementing the drills are an array of additional and supporting cutting tools, including carbide end mills, and linear motion products. Then there is the manufacturing equipment designed in-house, 100 percent in the case of PCB drills, 80 percent for carbide end mills, 60 percent for linear motion products, and an 85 percent average for all the company's products (Union Tool undated catalogue). Most notable here has been the development of the machine tools to enable the automation of the shrinkage fitting process. Such development depends on constant feedback and concentration of effort, not only within the company but also involving the customers. Proximity is therefore a key factor in serving the cream of the domestic electronics industry: Canon, Sony, Matsushita, TDK, Ricoh, Yokogawa Electric and many more. It likewise explains the placing of manufacturing bases in California, Taiwan and Shanghai where overseas customers are clustered. There is also a concentration of knowhow, however, in that all the essential machinery and equipment used by these overseas affiliates emanates from Niigata where the most advanced manufacturing is done and automation and rationalization have reduced the labor input to around ten percent of the production cost (JMAM, 2001).

Motion – articulation

When considering how Union Tool articulates to advantage it is worth bearing in mind that it is up against some pretty big rivals. In Japan, for example, they include the *keiretsu*-linked Toshiba Tungaloy, Mitsubishi Materials, and Sumitomo Electric. By comparison Union Tool is reminiscent of the Swiss mountain fighters encountered in Chapter 4. Both are compact dynamos confuting the odds. Reflecting this, Union Tool's articulation strategy is ultra-focus. A focus on technology, corroborated by a focus on cost and market, comprises the proactive toolbox of this articulation. The epicenter of the technology is the tungsten carbide drill bit represented by the bull's eye in Figure 9.1. The inner circle immediately surrounding the bull's eye is the shrinkage technology, together with the machine tools and accumulated skills involved in grooving, finishing and pointing. The outer circle constitutes the processing and other related machinery used in-house, but also for sale.

Within the ambit of the inner circle, the ability to adjust and realign the proprietary machine tools when new equipment is introduced is essential for ensuring the required quality and yield. Moreover, incremental upgrading of the machine tools themselves is a must for further product improvements. The fact that the drill manufacturing lines and the machine tool assembly lines are effectively adjacent to each other in Niigata renders such enabling feedback speedy and viable. Relying on outside servicing would slow down the capacity to respond, and in any case it would inevitably result in the leakage of knowhow to an undesirable degree in that if the competition got hold of it Union Tool's technological advantage could be compromised. So the inner circle is unpatented because the very survival of the company *in its present form* is almost exclusively dependent on its being impenetrable for outsiders. However, by directing so much of its attention to in-house provision of the means for realizing the finished product, Union Tool could be regarded as much as a machine tool maker as a PCB drill manufacturer. And this is where the link is formed with the outer circle, for Union Tool also sells machinery and devices it has developed which are deemed sufficiently distanced from its core knowhow. On the one hand, then, the conviction prevails that any untoward dependence on

Figure 9.1   The Three Basic Elements of Union Tool's Articulation

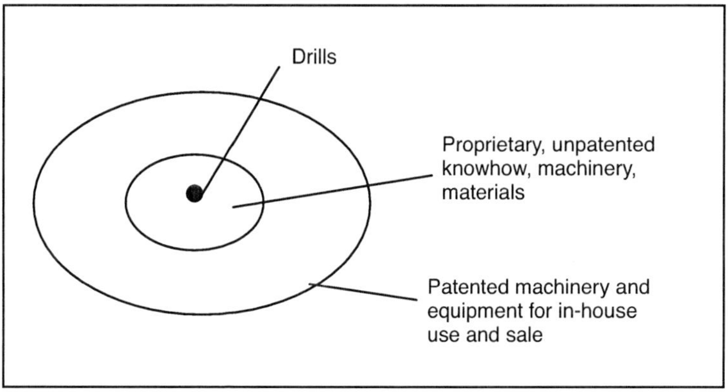

Source: Evans (2003).

outsiders could ultimately undermine the company's competitive position, while on the other hand there is the desire to broaden the base of operations in recognition of the vulnerability inherent in too narrow a concentration of effort on PCB drills.

However, such an eventuality is kept at bay by the other two supporting aspects of focus. For, although predicated on its technological focus, Union Tool's ultimate competitive advantage rests on cost which in turn feeds into customer service. For one, limiting the use of tungsten to the drill bit only while grafting on cheap steel for the shaft reduces Union Tool's materials costs to some two thirds of those of its rivals. Added to which, it is estimated that in-house production of machine tools and other intermediary manufacturing items reduces the company's equipment bill to one third of what it would be if it relied on outside suppliers (JMAM, 2001). Having the means of manufacturing immediately on hand, moreover, allows Union Tool to react to changes in demand faster and at lower additional overall expense than the competition. But such preoccupation with cost-cutting does not extend to labor because the market then takes precedence. Production is concentrated in the first instance in Japan for Japanese clients, in Taiwan for Taiwanese clients, in the United States for American clients. Unlike many other Japanese LMEs, Union Tool has never set up manufacturing operations in Southeast Asia to supply advanced-country markets. True, the labor card is played to an extent in China, but the affiliate plants are there first and foremost because PCB makers are there. Alternatively, being based at the doorstep of one of the crucial sources of information in California far outweighs in importance the burden of extra personnel costs.

Unity – the chamber

Union Tool functions on the principle that good *monozukuri* is based on good machinery. Its chamber encases a philosophy that insists on the in-house production of the machinery and other equipment for the manufacture of PCB drills, and it is from this that the company derives its strength. For from this philosophy is afforded the space in which internal activity can breath and move to its own rhythms. The knowhow is accumulated for making the machine tools, setting up the equipment, and responding to the feedback from the drill manufacturing line. This inward-looking focus is the better to prepare for dealings with the outside. In that reliance on outsourcing of equipment is minimized, the wellspring of innovation bespeaks the company's ethos. And there is that much less inconvenience with late deliveries, that much less bother with suppliers' home-grown solutions which do not quite gel with what the company feels it is about. Union Tool's chamber therefore embodies the structure which allows internal creative and reactive mechanisms to function while mediating contact with the exterior. With defense in place, the company is primed to attack with its unique and competitively priced drills and end mills and related equipment. This, in turn, is based on a strategy of product differentiation, backed by employee loyalty and close cooperation with the customer, with the intention of sustaining corporate autonomy.

But there is also another side to this particular chamber, and that is what is believed to stem from good Japanese-style *monozukuri* practices. President Katayama Takao is of the opinion that if such aspects as attitude to quality, the educational level of the plant operators, and awareness of belonging to the company are taken together, Japan is the least expensive in terms of personnel costs as a percentage of total production costs. Where, as is often the case outside Japan so he believes, quality is not given the attention it deserves, far from savings being made, the costs mount up. This being so, sustained added value is more assured in Japan and there is no intention on Union Tool's part of shifting operations abroad beyond what is deemed absolutely necessary as dictated by the need for market proximity (JMAM, 2001:236). In other words, there is no soft utopianism in Union Tool's chamber. It is designed for autonomy and autonomy is most securely achieved by earning power. This is practical productivism predicated on full commitment to *monozukuri*. It is an approach which has resulted in hefty net profit increases over the past few years, and is reflected in its healthy financial position as indicated in Table 9.3.

Table 9.3   Comparison of Financial Results for Union Tool and Other Listed Manufacturers (based on FY November 2007)

	Union Tool	Average
Sales growth	+5.9%	+9.4%
Operating income growth	−3.9%	+10.0%
OP margin ratio	26.1%	6.7%
Return on equity ratio	12.0%	9.4%
Shareholders' equity to total assets	89.7%	41.2%

Source: http://www.uniontool.co.jp/english/pdf/faq_e_08011.pdf
NB: 'Average' means the financial index average for manufacturing companies listed on the Tokyo Stock Exchange.

Summary

Taken on average the Japanese LME still retains four distinctive features. First, it spans the ownership/managerial divide, often attached to varying degrees to the founder family. Second, its core company within Japan has remained stable in terms of number of employees. Third, it acts independently and is not dominated by larger corporations. Fourth, it contributes to the Japanese economy's 'full set'. However, the economic environment has changed considerably since the 1970s and this has given rise to increasingly varied responses by LMEs, more and more of which can be associated with external influences. Such responses are represented by way of three profiles.

With its American-style approach, Nidec has grown big by acquiring and absorbing other companies both in Japan and overseas. This fundamental attitude stems from the experience and reactions of its maverick founder and president, Nagamori Shigenobu, who started the company when he was 27 years old and forced its presence on the nation's samurai business world initially by a 'reverse landing' via the United States. Nidec is also atypical in Japan, or perhaps merely ahead of its times, in the aggressive emphasis it places on money, both in terms of profits and employee remuneration. Takara Tomy, on the other hand, directs attention to the adjustments Japanese LMEs are increasingly being called upon to make due to changes in circumstances. The act of merger of the two companies in the same industry and around the same size is reminiscent of the German practice over some decades. Having said that, the follow through involving additional tie-ups and the co-opting of an investment fund's expertise for international marketing is decidedly American.

Nevertheless, ultra-focused specialization Japanese-style is still viable in the modern setting. A theme throughout the book has been that there is a fit between the Japanese LME and the stylized LME depicted in Chapter 4. By way of illustration, Union Tool, which evinces marked similarities with this stylized LME, is matched against it using the same headings: 'Entity – size, stage and type', 'Evolution – background and process', 'Motion – articulation', and 'Unity – the chamber'. The company's core competence is centered on the tungsten carbide drill used for making minute holes in PCB boards. Since starting this business in the 1970s, it has become one of the world's largest producers and highly profitable. This success is presaged on the concentration of proprietary know-how and an intensely targeted organizational structure.

10
Monozukuri – Making Things

The essence

The three key determinants shaping the Japanese economy and the LME within it discussed in Chapter 8 can be lodged within the realm of the 'samurai'. It was the upper echelons of the national hierarchy who determined the absolute necessity of redemption and restitution, now to be realized through the avenues of economic progress. It was the central bureaucracy which shouldered the responsibility for arbitraging with global institutions to mitigate as much as possible the repercussions of demanding external forces, while at the same time devising measures to reinforce the economy and the component organizations within it in readiness for what was to come. But there is another determinant, and its essence resides with the 'artisan'. It is *monozukuri*, or making things. *Monozukuri* purportedly implies for the Japanese something which is commonly and uniquely shared and mutually understood. In their eyes *monozukuri* acquires a particular essence which is quintessentially Japanese, just like their home-grown glutinous rice, whose sanctity renders price all but irrelevant. In its pure form the *monozukuri* mentality puts the making of things first; this is where the heart of the attention and concentration of the enterprise lies. Its indispensable stakeholder is the creator-engineer, the artisan. Ignore the artisan and you ignore *monozukuri* and how it is embedded in the firm as the fundamental driver of production.

As such, needless to say, unlike the other three determinants its roots go way back beyond the modern era; it is far from being simply postwar in origin. However, over recent years, in the hands of the central bureaucracy the term *monozukuri* has been co-opted and broadened in its scope of reference to include organizational efficiency in business activity all round. Put simply, it has been modernized. In one sense pristine *monozukuri* has been usurped by the samurai for his own urbane ends. On the other hand, though, we can say that it has come to serve as a collective national totem, notwithstanding the diverging emphases ascribed to it. Indeed, what could be more symbolic of such a national consensus than the fact that in 2005, Fukuda Metal Foil & Powder, founded in 1700 be it recalled, was the proud recipient of the prime minister's first 'Monozukuri Nippon Grand Award'?

Monozukuri, then, is something that has germinated and grown within the nation's collective consciousness. Taken back to its essence and shorn of ensuing accretions it can be brought to life with the following example. Asano Taiko makes traditional Japanese drums, which is what 'taiko' means. It has been making them for well nigh 400 years, having been founded in the Kaga domain, now part of Ishikawa Precture overlooking the Sea of Japan. The actual year of the company's founding was 1609, nine years after the Battle of Sekigahara in 1600, the 1066 or 1776 of every Japanese schoolchild, which ushered in the Tokugawa era. Nowadays, with its 20-odd staff of craftsmen and assistants, Asano produces about 8000 drums a year for a domestic market share of around 70 percent, in an integrated sequence from felling the trees, boring out the various shapes and lacquering the wood, to processing the leather and stretching it over the finished item. For this, even now into the 21st century, it is winning national awards for good design (www.asano.jp).

Moreover, Asano Taiko neatly embraces some cardinal aspects of our story. It has existed throughout the extended time frame of our narrative, while being founded and remaining ever since in a place relatively remote from the nation's industrial hubs, thereby evidencing the relative dispersion of Japan's economic activity and, as a corollary, its *monozukuri*. Further, Asano has been extraordinarily consistent in its chosen vocation, and it produces something intrinsically Japanese – still measured in *shaku* and *sun*, non-metric as ever – while answering to a 'modern' market demand. Not that being intrinsically Japanese bars others from *monozukuri*, needless to say. But as already indicated the Japanese have, in their eyes at least, arrogated it to themselves and given it special significance in their culture. How valid or justified this is not the issue. The fact is that believing that it is part of them influences the way they think and act. Awareness of it has also led to progressive embellishment. Asano Taiko evokes the primal essence; the artisan has worked to its rhythms; not to be left out, the latter-day samurai has co-opted it; and the LME has emanated from it.

The artisan's *monozukuri*

The discussion pursued by a couple of veterans of the cause as recorded in a recent book can introduce us to the artisan's perspective (Matsuura and Okano, 2004). True to this genre of book in Japan, these gentlemen come across as opinionated, irascible, iconoclastic and a tad xenophobic and could be dismissed as such, but for one thing: they are both possessed of impeccable *monozukuri* accreditation with international reputations. Okano Masayuki is a world-class craftsman who has attracted the attention of major multinationals and NASA for his metalworking genius. Among his recent exploits, this septuagenarian with a complement of six or seven employees occupying a workshop in downtown Tokyo, has produced the world's thinnest needle for incorporation into a painless syringe. The needle constitutes a cone of sheet metal so thin that it was deemed by everybody else impossible to roll up. But then Mr. Okano proved otherwise. Thus is revealed the passion of the true artisan working against all odds to challenge the impossible. Matsuura Motoo, on the other hand and by his own admission, is not so much an artisan but rather a company founder. Nevertheless, in instigating the production, by his company Juken Kogyo, of the world's smallest gear, weighing just one millionth of a gram with a diameter of 0.1489 mm, he has likewise probed the outer reaches of *monozukuri*. And these extremities can go beyond the realms of the 'practical'. Nobody asked what such a gear could be used for while it was being developed; it was simply there to be accomplished. *However*, once accomplished, it has served as a bait, as Mr. Matsuura calls it, because it has activated lucrative orders for miniaturized items from clients around the globe intrigued by the capabilities thus demonstrated at Juken Kogyo.

A Japanese domain ...

Be that as it may, starting at the purely idealistic end of the *monozukuri* spectrum, a firm, according to Matsuura Motoo is a crystallization of your intent and ideals, epitomized in the development of ultra-small gears for example. Place that within an unapologetically nationalist frame and the spirit of *monozukuri* is embodied in the artisans of Edo; have-craft-will-travel, penniless in the morning but blessed with the skilled hands that feed. It is the Japanese sword assiduously crafted, to be kept as a family heirloom gracing the *tokonoma* (alcove), not so much for use or sale, but to satisfy a yearning for ultimate beauty. What makes the character of the Japanese come alive is to be found in nothing other than *monozukuri*; Japanese culture *resides* in *monozukuri*.

This is also expressed in what are perceived as peculiarly national attitudes towards time, approach and organization as they pertain to *monozukuri*. Simply working the machine is not enough; repeated operation is needed to acquire a delicate touch. It is also necessary to understand just by looking at the material and the drawings what the problems in the processing are and judging what processing method would be suitable and effective. Getting all this right can take at least five years for, say, molding, casting and metal cutting (Fuji Research Institute, 1998:31–2). But as he progresses along this path the operator has his interest and curiosity aroused so that he will quite happily engage himself in maintenance and repair of the machine, something non-Japanese operators would regard as beyond their remit, so the story goes. Received attitudes towards *monozukuri* also posit that operator and his company in a particularist environment, often entailing long-term loyalty of a highly specialist nature to a reliable clientele. At its best, moreover, the *monozukuri* mind embraces the whole company; the humble operator is ostensibly joined by everyone from the foreman to the CEO in a mutual concern for the minutest detail. Indeed, Toyota of all companies presents a perfect illustration with its practice of *genchi genbutsu*, loosely rendered as 'go and see for yourself', which sends top executives on regular forays to the shop floor (Liker, 2004).

But our two purists voice pessimism. There is a loss of feel for the métier, they say, as symbolized by the fate of the *shaku* and *sun*, organic and tangible to the Japanese experience, where Asano Taiko represents an anomaly being progressively crowded out by the tyranny of a metric system – cold, indifferent, and not invented here. The *shaku* and *sun*, after all, once undergirded native creativity. But whereas in the past standards emerged *de facto* from the fertile soil of practice and perfection organically evolving from traditional foundations, they are now being imposed *de jure* by external regulators, both domestic and increasingly international. Looming ominously among the latter is ISO. Matsuura Motoo's company, Juken Kogyo, has not applied for ISO because he thinks it is irrelevant. In his view it is simply a means of erecting the supposed right management system. It is incompatible with Japan's *monozukuri* approach which entails first and foremost an intense examination of how to go about making something good. Moreover, because it is an external imposition targeting incidental organization as opposed to core purpose of the manufacturing entity, it stifles inherent ingenuity. Being sacrificed in the process is home-grown total quality control (TQC) (Matsuura and Okano, 2004:23–7).

... amplified by outside inputs

However, before we get too bogged down in this lugubrious brew of patriotic idealization and gloom, we need to take stock of the broader picture. To take the obvious first, and as our two protagonists would freely admit, *monozukuri* is not the sole preserve of the Japanese: the tales already related of Andreas Stihl and his powered chain saw, Earl Bakken and his pacemaker, as well as many, many others are proof to the contrary. Come to that, American executives at Toyota's American-based plants are apparently eager practitioners of seeing for themselves and getting their hands dirty. More open to question, though, is the impression that TQC is somehow a Japanese invention. It is not. TQC derives from the discipline of Scientific Management, as originally formulated by mechanical engineer Frederick Winslow Taylor, hence Taylorism. The point rather is that the Japanese were arguably the world's most zealous converts to Scientific Management as it subsequently developed during the 20th century. This arrived at a juncture around the second decade after the Second World War when it had become so ingrained in business thought and practice that, as their own brand of what became known as TQC was perfected, many Japanese assumed it to be of all their own making.

That is as may be, but on the other hand, on the road to TQC there can be little doubt that Japanese *monozukuri* notions were liberally inserted. In other words, Scientific Management was yet another example of Japanese adaptation, initially perceived through an optic somewhat askew to that of the American conception, then over time dosed with native instincts and preferences. So this brief clarification does not make the Japanese any less justified in believing in their exceptional inclination and aptitude for *monozukuri*; it merely seeks to balance the scales in the telling. Scientific Management, moreover, was far from being merely the denizen of the shop floor, especially in corporate America. It was an intellectual exercise professing universal principles to accommodate the age of mass production, the worthy successor to the creed of bonding between sword-smith and apprentice. As such it engaged the attention of those ultimately concerned with the implications of *monozukuri* on a national scale: the samurai bureaucracy.

The samurai's *monozukuri*

Not that *monozukuri* was uttered in the same breath as Taylorism in the early 20th century needless to say, or at least not with the degree of meaningful content it now evokes. In fact, from the bureaucratic samurai's perspective, *monozukuri* is quite a recent descant added to the tune of resurgent industrial policy. When we last left Japanese industrial policy – or perhaps more accurately now described as the (perpetuated) administrative stewardship over the broad outlines of national economic sustainment and renewal – it was caught up in the consequences of economic success, buffeted by international pressures and intent on modernizing the unmodernized in response. However, from the 1960s on, the main instigator of industrial policy at the time, the Ministry of International Trade and Industry (MITI), was progressively forfeiting its direct influence over foreign exchange, credit, and resource allocation. Larger corporations had to a considerable extent outgrown the need for government attention and largesse, while by the 1980s changes in the economic structure and how it functioned were driving home the fact that small firms came in many stripes, not a few of them vital and perfectly capable of standing on their own feet. In this new environment, policymaking took on a more modest persona, not so much guiding as assisting in, for instance, the realization of technological development and the diversification of depressed industries. All this kept time with a new buzzword for each decade, from modernization in the 1960s to knowledge intensification in the 1970s, information technology in the 1980s, industrial sector fusion in the 1990s.

The administration, however, was still actively involved. The end of 1995 saw the promulgation of the Science and Technology Basic Law, which stressed the essential role science and technology should play in the development of the Japanese economy and society, instigating reforms aimed at greater financial input for R&D, more exchanges between the private sector and the university, more emphasis to be placed on S&T education, and furthering international cooperation in research. Moreover, MITI – the chief samurai of industrial administration and predecessor of the Ministry of Economy, Trade and Industry (METI) which was to take over in 2001 – still retained an invaluable trump card: its pivotal nodal position (Fransman, 1999:162). It was at the center of an information network of national and now global proportions, strategically placed like none other to fashion its own version of *monozukuri*.

The official stamp and reformulation

And *monozukuri* was the word employed. In 2001, with the implementation of an administrative reform program, MITI became METI, and with that METI in collaboration with the likewise reconstituted Ministries of Health, Labor and Welfare and of Education and Science launched a new annual report, called in Japanese the *Monozukuri Hakusho*. This can be literally translated as 'the white paper on making things'. However, its official English title is the 'White Paper on Manufacturing Infrastructure', no doubt on the grounds that the literal translation would not impart the gravity of the undertaking to non-Japanese, who in any case would be unlikely to appreciate the responsive chord struck among the Japanese by such an ostensibly lame expression as 'making things'. In fact, on the surface, the white paper is just what the English rendition says it is, a report on developments nationally and globally pertinent to Japanese manufacturing and an outline of measures, ongoing and desirable, to address the evolving situation.

But there is also a premise involved which shapes the approach as, indeed, enunciated by the white paper itself. In the past, it says, what was viewed as important among those espousing *monozukuri* was the skill of the artisan. However, it claims that by the 1980s the baton of *monozukuri* had been passed to the computer-conversant technician. This in turn, rather reminiscent of how marketing a few decades earlier had in the eyes of its advocates engulfed sales, procurement, production and what have you and subordinated them to its ultimate direction and control, has led theorists in the administration to present *monozukuri* as an all-encompassing emblem of excellence, even as a virtual panacea. Hence, this version of the *monozukuri* man is multi-skilled and adept at operational planning; he understands the needs of the market, is capable of absorbing information at a pace, and stands ready to make constructive suggestions regarding design and development; he is an imaginative leader motivating his subordinates (METI *et al*, 2004:392). Further still, by converting *monozukuri* to an integrated system with information technology as the axis, there will be a reduced need for readjustments resulting from experimental and design errors when employing complex mechanisms, the time required for manufacture will be reduced, and R&D will be carried out to build information systems for enhancing new technology and production development (METI *et al*, 2004:443). Thus has the urbane samurai assumed the native garb, but in satin not cotton.

There is ample reason for the interest in manufacturing, needless to say. Simply put, manufacturing remains the linchpin of Japan's economy. To take some pointers from 2002 by way of illustration. Manufacturing accounted for 22.6 percent of value added within GDP, second only to Germany among the world's major economies. Ninety percent of Japan's exports for that year were industrial manufactured goods, confirming the role of manufacturing as the nation's chief procurer of foreign exchange. Nearly 20 percent of the active workforce owed their jobs to manufacturing and the manufacturing industries bear the brunt of private sector R&D expenditure, second again in both cases only to Germany and ahead of the United States, Britain and France. In fact the status of manufacturing has if anything been bolstered over recent years. Since 1990, right through the prolonged recession be it noted, labor productivity among the manufacturing industries has substantially exceeded that for industry as a whole to make manufacturing the motive power for such growth that has occurred (METI *et al*, 2004:27–32). More significant still, in reversing the hollowing-out tendency, Japan's manufacturing is now confuting received globalization logic. According to METI's 2003 edition of a periodic trends survey of industrial location, there was an increase in the number of new factories opening up domestically. Manufacturing was coming home from overseas plants, suggesting that problems associated with domestic production had been overcome and that advantages were perceived anew in this, notably with respect to quality and supply chain management, which in turn strengthened international competitiveness (METI *et al*, 2004:216).

The official contribution

The administration, as represented by the three white paper ministries here, for its part, while actively encouraging relocation back home does not regard this return to the roots as in any way a disengagement from globalization. Quite the contrary, it is strategy aimed at more effective *engagement*. Indeed, it sees one of its roles as an interface with the global arena and as such endeavoring to ensure that international standards are met. Another essential role for government, as perceived by the administration, is as architect of a broad vision which advocates as priority targets at this juncture life sciences, information and communications, the environment, and nanotechnology materials. Then at the ground level it becomes the orchestrator and expediter of institutional collaboration, guiding university research to practical ends and private sector funding, linking venture capitalists to organizations with technology to transfer, devising regional industrial cluster plans involving broadly-based industries, academe, national and local government networks, LMEs and SMEs.

Complementing this is the establishment of consummating institutions, some with unequivocal elitist coloring like the Monotsukuri – different spelling, same meaning – University, or the Institute of Technologies to give it its official English title, in Saitama Prefecture. Founded in 1999 and currently headed by the chairman of Hitachi, this select Institute admitting only some 250 students per year provides a full four-year university course aimed at producing the 'monozukuri leader' possessed of the all-round capabilities trumpeted by the Monozukuri White Paper. Hence, the samurai's *monozukuri* embraces all: the national administration, the regional association, the company, and the individual. *Monozukuri* has waxed mightily from the artisan's original simple prescription. How and to what extent this spectrum has influenced the LME's progress in postwar Japan, given the other three determinants of growth, industrial planning, and international pressures, is the subject of the following section.

Evolving *monozukuri* and the LME

The stylized LME presented in Chapter 4 should leave the reader in little doubt that *monozukuri*, as manifested in its various aspects, is regarded here as an integral part of the LME's makeup. Here it is proposed to draw together the essentials as represented by the individual, the firm and the institutional environment in an attempt to summarize the evolution of the broad thrust of *monozukuri* up to the present. In other words, this exercise takes us again along the artisan-samurai spectrum, as well as the time sequence we have been following from the beginning of the Meiji era and Japan's industrialization to the early years of the new millennium.

From the individual to the firm

The point to bear in mind, of course, when contemplating Japan's industrial pioneers is that *monozukuri* is not the equivalent of invention. It is not even a synonym for innovation, although there is often a fair amount of that around. First and foremost *monozukuri*, especially for the individual artisan, is a state of mind – curiosity, ambition, tenacity – which impels action. This is why so many Japanese did not assume a posture of indifference when the West appeared at the front gate with its novelties; they walked towards them to inspect, to evaluate them, and to appreciate the whisky. But although trading held its appeal, it was the *making* that most intrigued.

True, government prodding did induce outcomes in many cases; that is how Koransha's Fukagawa Eizaemon VIII applied his ceramic expertise to insulators and conceivably how Tanaka Umekichi was pointed in the direction of platinum. But it is highly unlikely that S&B Food's

Yamazaki Minejiro was chaperoned by a government agent on his trysts with curry in downtown Tokyo. Shimano Shozaburo's *monozukuri* – likewise self-inspired – proved to be a classic instance of incrementalism, adapting product and process, both originating abroad, so that they would ultimately take on the world. 'Carbonated paper' had first appeared while Napoleon was sweeping across continental Europe in the early 19th century, although it was not until 1880 that Gestetner's Cyclostyle was patented. So Ashida Junosuke, founder of General, picked up on carbon paper at the consummate moment when he could combine the imported concept with the ink he himself contrived and paper long since in production in neighboring Kawanoe. Okano Masayuki, perfecting his superthin needle at the start of the 21st century, is a direct descendant of this spirit. Moreover, it might be added that these are not illustrations of the *hitozukuri* of the samurai's drawing board; these people *made themselves* to make the things.

The fact is that there is no time that the self-motivated individual departs from the *monozukuri* scene. As already noted, the authority of the individual is writ large on Japan's economic success. Alternatively, though, *monozukuri* did not live or die depending purely on the attentions of the outstanding individual. It infiltrated the communal, business and institutional environment. Shofu Katei provides an example of this in comparatively early industrializing times with his establishment of the Shofu Porcelain Teeth Research Institute, thereby facilitating and promoting the spread of the *monozukuri* ethos among a team of in-house researchers linking up with reputed outside specialists. Staff and co-opted expertise were part of the act; the firm, in turn, was a node of *monozukuri* activity. As the prewar economy became larger and more complex new firms, often very small ones, joined the fray, forming the foundations of a common commitment. The *byora* makers of Higashi-Osaka elaborated their turf as machine makers arrived, attracted by the fact that the *byora* makers were already there. Such concentrations accelerated the intersecting of the economic components, at the same time setting boundaries of specialization which – even for the very small casting or heat treatment workshops – had the dual effect of reinforcing positions while constraining entry. Firms were afforded space to hone their expertise; the creed of *monozukuri* generated core competence, while for many, scope took precedence over scale.

Monozukuri and networking

By the 1960s, then, Japan had developed a highly distinctive industrial structure taking the form of widely divergent manufacturing companies with the large corporations at the apex. It existed on the basis of advanced production and processing technologies accumulated by these corporations coupled with thick layers of LMEs and SMEs providing the sites for making components, materials, machine tools and so on. In manufacturing said items these latter companies combined in networks with smaller specialists often within geographical proximity. This networking aspect has been further elaborated since then so that, by way of illustration, the machinery industry has been likened to a mountain range. Peaks representing the large corporations still protrude, but as one descends to the broad foothills it is no longer possible to associate where one is standing with any one of a number of peaks. Down here the small-scale processor with his variegated client base is not stuck at the end of a highway leading up inexorably to one particular peak, but occupies a dip in a valley where many paths criss-cross (Fuji Research Institute, 1998:8). In this field of *monozukuri* give-and-take the large assembler has come to rely on the supplier as much as the obverse. This de-pyramidalizing – if one can be forgiven such a concoction – affords scope for firms of all sizes.

It has already been noted that LMEs have been able to amplify their status by availing themselves of the services of SME specialists. But an SME, no less, can be cited to drive this point home. Morico Co., Ltd. makes stamp canceling machines. Having initially started by supplying these machines to the postal service, its business grew fast from the middle of the 1960s subsequent to the passage of a series of legislation which obliged food, pharmaceutical and other manufacturers to stamp the date of production on their packages. Although Morico once had a larger workforce, it has since settled at around 55, with total gross sales for 1998 of seven to eight million dollars. Business stability is sustained by combining extensive proprietary know-how accumulated over decades, concerning the most effective materials and synchronizing techniques, with networking, both for supply and demand. It has on call no less than 100 subcontractors, half of whom have been set up by former employees. At the other end of the chain its clientele reads like a who's who of the food, chemical, pharmaceutical and cosmetics industries as well as covering machinery, electrical and electronics, precision equipment, transportation, trading and other companies, plus the post office which still accounts for some 40 percent of Morico's business (www.morico.co.jp; Hibi, 2002:102–10). It is in this kind of world Japanese LMEs grew during the postwar and high-growth decades.

The developing LME advantage

Indeed, many LMEs still flourish in it. In fact, since the 1970s and more so since the 1980s the tectonic plates of economic structure and practice in Japan have been gouging out a transformation, much of it favorable to the LME cause. As of the 1970s the industrial structure, not only of large corporations but also of many smaller firms was changing to one of high value added, this being particularly the case for metal products as well as general, electrical, transportation and precision machinery for instance, which were coming to require an advanced degree of processing (Ito, 1996: 50–1). This was the start of a progression whereby structure was increasingly called upon to accommodate evolving priorities including the marshalling of information, post-industrial technologies and ultimately the demands of globalization. This structural transformation was accompanied by industrial changes such as intensified market segmentation and sectoral merging or fusion, along with the emerging pre-eminence of economies of scope and networking.

The implications of these developments can be seen with respect to both production and management. To take metal products to illustrate the production aspect, although mold making still requires at the top of the range high degrees of technical skill only possessed by the specialist, it is now possible to mechanize practically all the processes where reliance on manual skills is small, thereby encouraging labor-saving through automation. Added to which, with the slowdown in domestic automobile manufacturing in the mid-1990s there were fewer orders for castings. There again, in the area of electronics, specifically personal computers and mobile phones for example, with the progress in making devices thinner and lighter, there is a demand for parts of ever higher accuracy and precision. In all three cases the outcome has been the same: the smaller operator has lost out to the larger firm.

The backstreet mold maker cannot afford the necessary array of numerically controlled machine tools and machining centers which enable high-speed processing to meet short delivery deadlines; the expensive, complex equipment is likewise beyond the reach of his metal stamping confrère (Fuji Research Institute, 1998). Moreover, when demand has contracted, producers have tended to rationalize supply chains, which has usually favored the larger supplier. What is the small operator's loss is the LME's gain; superior financial arrangements being one reason for this, management capability being another. LMEs have built up their design and R&D departments and having thus strengthened and elaborated their functions have positioned themselves as increasingly indispensable *partners* with the large assemblers. At the same time, their more advanced and flexible understanding of strategy has enabled them to be actively engaged in diversifying economic activity as industrial sectors increasingly merge and interact. The samurai's *monozukuri* takes shape, at one and the same time more extended in scope and more concentrated in location.

A convergence of interests

For it is here that all the key references we have been discussing come together. In the first instance, it is where the administrative perception of Japan's *monozukuri* potential in a globalizing environment and where that potential demonstrated by LMEs are in the process of converging and where, therefore, the administration and the LME share interests in common. In this convergence scenario the Japanese LME still remains true to itself; it is intent upon constantly clarifying the skills and knowhow comprising its own strengths and core competences. At the same time, however, the LME is increasingly seeking external support and stimulation. For one, as reported in a recent survey, it expects of the state the provision of a practical educational and training system, that is to say the facilities on a national basis for *hitozukuri* – making people (METI *et al*, 2004:393). To an extent this implies the desire for a more universal, portable skill base, allowing for more workforce mobility and therefore more flexible hiring and firing practices.

The LME also concurs with the need to adapt to the globalizing exigencies, thus implying that the stylized LME is on the move in response to maturing circumstances. This calls for fusion, both between *monozukuri* and information technology and, beyond the LME's stylized chamber, with other manufacturers in the same and different industries, production designers, marketing specialists and what have you. All of which ties in with the administrator's vision and the choices globalization offers predicated on Japan's accomplishments, a vision also shared by many in business as these choices become clearer.

Simply put it states that Japan's *hitozukuri* and *monozukuri* drive its globalization strategy. Cheap factors of production – notably labor – abroad have their place, but if, as is often the case with Japanese firms, the product demands extremely high levels of precision which can only be rendered under the watchful eyes of well-educated, meticulously trained, disciplined operators and technicians, then it is more cost-effective overall to produce in Japan – for the time being. Aggregate this talent and it amounts to a large number of manufacturers with a multitude of technologies at their disposal covering materials, parts, capital goods, assembly and processing, in what could be regarded as a vast clustering of nationwide proportions, which is to boot adjacent to a rich and highly receptive home market. This has been the substance over recent years of the government's argument for keeping or returning production to Japan; it also fits the pattern of business, in part or in total, of many LMEs as employers, suppliers, procurers and networkers. It does not preclude FDI and other forms of overseas presence, needless to say. Rather, it constitutes the basis of a more refined international division of labor which modifies the approach to globalization that LMEs are taking subject to no small extent to the *monozukuri* translated into core competences they have on offer. Moreover, it can be seen as a reassessment of the country's competitive advantages in which, taking into account the reality of new emerging forces – most notably China, Japan's positive strengths in terms of factor conditions (*monozukuri*), demand conditions (high-grade intermediary and finished products), related and supporting industries (as represented by LMEs), and firm positioning (networking and rivalry) are emphasized (Porter, 1985). And all this still calls for people like Horiba Masao.

Horiba Masao – the modern *monozukuri* man

In 2006 Japan's pre-eminent economic daily, the Nihon Keizai Shinbun, or Nikkei, ran a short autobiographical series penned by the octogenarian Horiba Masao, founder and now *éminence grise* of Horiba Ltd., currently boasting an 80 percent world market share in engine emission analyzers (JCH, 2005). The title heading for the first article read, 'The radio receiver was the source of my *monozukuri*', and the first word of the first sentence was likewise *monozukuri*. This *monozukuri* man embodies all the elements of the spectrum that the concept has come to signify, from technological and scientific expertise, entrepreneurial pugnacity, institutional involvement and, finally, global perspective.

The technical *monozukuri* man

The technical man started in short trousers, in fact. Having acquired a taste for taking things apart and putting them together again at an early age, by the time he was at high school he had joined the radio club there and was occupying himself scouring the specialist outlets in Tokyo and Osaka for components which went towards, among other things, producing the power circuit for his very own short-wave receiver (Nikkei, 2006b). Thereafter, he studied physics at Kyoto University. He was writing his doctoral thesis when the Second World War came to an end and American Occupation policy dictated that the university's key nuclear physics experimental equipment be destroyed. Foiled in his desire to stay on as a university researcher, on 17 October 1945 he set up the Horiba Radio Laboratory in a two-story wooden building with a floor space of 80 square meters, where he worked while also completing his thesis. In a manner reminiscent of Medtronic's Earl Bakken, while making and repairing measuring equipment, he developed a storage battery for use in the event of power failures.

However, Horiba's ambitions for his laboratory were far grander. With the money he made from the battery he and his team embarked on the development of an electrolytic capacitor. This capacitor had better control qualities than other products at the time and was well received by large potential customers, so a factory for commercial manufacture was on the cards. But fate struck once more, this time in the form of skyrocketing inflation with the start of the Korean War, which trebled the costs entailed and disenchanted those potential customers when approached for funds. Nevertheless, there was a silver lining. In developing the capacitor, the laboratory team had found it necessary to devise Japan's first electrode-type pH meter to ensure the pH consistency of the solution used for making oxide film. Unlike electrolytic capacitors, substantial capital is not required to manufacture pH meters. So on 26 January 1953 a company – Horiba Ltd. – was established to do just that. Horiba Masao was then 29 years old (www.horiba.co.jp).

The company *monozukuri* man

Apart from his devotion to science Horiba Masao had indulged in one other passion during his university days: rugby. From this game he had absorbed a lesson which has ever afterwards been foremost in his mind. This was a commitment to teamwork. If the individual insisted on hogging the ball, both he and the team would ultimately lose out. Likewise, *monozukuri* had to be a team game and this was how the company was to be run (Nikkei, 2006b). A number of consequences flowed from this approach. First, reflecting the eventual adoption of the corporate philosophy of 'Joy and Fun' (*omoshiro-okashiku*), Horiba signaled to those who worked *with* him in the team that they were to give full expression to their own capabilities in a thoroughly amenable environment (Horiba, 2003). But you still want a team with the best players around, so they had to be good, too. They had to be able to do things others were not capable of, and if they could not live up to this they had to leave (Ishikawa and Nejo, 1999:116). Second, Horiba Masao himself acted *within* that team as a *member*. From the start he had succeeded in attracting highly capable graduates. However, after a time they voiced dissatisfaction that they were not presented with the same opportunities as their former fellows students who had entered large corporations. The small size of Horiba Ltd. did not allow for the same leeway, but nevertheless he encouraged his staff to take doctorates and, more to the point, joined in by taking a doctoral degree in medicine to show the way.

The third consequence was that he gained from the teamwork approach not only in having people work harmoniously but also in himself adhering to where the general mood of the team was taking the company. Having succeeded with the pH meter, the company began research on infrared measurement for use in analysis of gases. This was to lead by the end of the 1960s to the development of an emission gas analyzer for automobiles. Horiba Masao had been dubious about this particular application all along (Horiba, 2005). However, he did well to listen; those engine emission analyzers now account for 40 percent of the Horiba Ltd.'s business (Nikkei, 2005b). Finally, the 'team' has now been taken beyond the company's bounds. This in a sense puts a dent in our hypothesis of the stylized LME as a chamber, reminding us that hypotheses, like firms, must change with the times, especially nowadays with the Internet. In October 1997, Horiba's website came out with a new section called Advanced Business Collaboration or ABC, which presents products in the course of gestation to obtain customer feedback *ex ante*, thereby integrating actual needs with creative ideas. It is also one means for sustaining what the current president, Horiba Atsushi, calls 'a great medium-sized company', which after all in our terms is an LME (Ishikawa and Nejo, 1999: 71–2, 113).

The institutional *monozukuri* man

Like Shimano Shozaburo and Shofu Katei III before him, Horiba Masao has always relished the inclinations of the 'artisan'. However, at the same time he was never without 'samurai' connections, which is evident in the course of his early life. Son of a professor of chemistry at Kyoto Imperial University, as it was then, he graduated and received a doctorate from the same university (he was awarded his later medical doctorate by what is now Kobe University). Strapped for capital having developed the pH meter he could turn for funding to the Kyoto business community to establish Horiba Ltd. While chairman of the company from 1978 to 2005 he received several awards from the Japanese government, including a national Blue Ribbon Medal. That having been said, though, he has been very active in the local business community, notably as special advisor to the Advanced Software Technology and

Mechatronics Research Institute of Kyoto (ASTEM), one of the largest start-up incubator organizations in Japan, helping entrepreneurs set up durable, effective companies. At the national level, since its inauguration in 1999 he has served as a representative of the Japan Association of New Business Incubation Organizations (JANBO) (Chemical Heritage Foundation, 2005). This association has established one-stop service centers in each of Japan's 47 prefectures, which in Horiba's eyes can function as hubs for revitalizing local industries, thereby readjusting a national structure at the moment too concentrated on the great metropolises – an artisan's response to the samurai's excessive organizational urge perhaps.

The global monozukuri man

And now, not far shy of 40 years after the emission gas analyzer hit the American shores, Horiba Masao is enjoying global recognition. On 4 November 2005 it was announced that he would receive the Pittcon Heritage Award, thereby being the first recipient from Asia and only the second non-American. It is presented to outstanding individuals whose technological and entrepreneurial flair has contributed to spreading the word of instrumentation science and analytical chemistry to a worldwide audience (Chemical Heritage Foundation, 2005). Understandably Dr. Horiba was delighted to receive such international acclaim after 60 years of effort. But he also has a global message to impart stemming from his experience. Being also president of the Association of Asian Business Incubation (AABI), directly inspired by JANBO, in a recent keynote address to AABI, Dr. Horiba expressed his hope that Asian countries would extricate themselves from the subcontracting mold to develop local industries of their own unique provenance in the same way that JANBO is advocating for Japan's regions. The world economy, he continues, has been terribly affected by terrorist activity. The reasons for such actions are many and complex but underneath it all is the wealth gap. Every effort must be made to close this gap throughout the world (Horiba, 2002). *Monozukuri*, then, is viewed by this modern *monozukuri* man as an essential contributor to global prosperity and peace.

Summary

The fourth determinant of Japan's postwar development, but with a much longer history, is *monozukuri*, or 'making things'. Its essence resides with the artisan although it has now been adopted by the samurai and broadened in scope to become a national totem, symbolized by Asano

Taiko's drums. The artisan's *monozukuri* is personified in Okano Masayuki, maker of the world's thinnest needle, and Matsuura Motoo, instigator of the world's smallest gear. To them *monozukuri* is represented by the Japanese sword which satisfies the desire for ultimate beauty. *Monozukuri* is embodied in the conscientious attitude of the machine operator and adhered to by the whole company. The devotion of the Japanese to *monozukuri* has also turned them into the most avid practitioners of Scientific Management and TQC.

Moreover, there is now samurai *monozukuri* which found expression with the bureaucracy's need to redefine its role as adviser and prompter as it looked to highlight a succession of priorities: modernization, knowledge intensification, information technology, sector fusion. It began to take definite shape with the arrival of METI and the publication of the *monozukuri* white paper with the advent of the new millennium. Here *monozukuri* is no longer epitomized by the skill of the artisan, but by the modern multi-skilled, market-savvy, motivating leader. It is an integrated system based on information technology. The reasons for stressing it are that manufacturing is still the linchpin of the Japanese economy and if anything is seeing a resurgence with production being repatriated to an extent. To encourage such an eventuality is not seen by the administration as a disengagement with globalization, but on the contrary a more effective means of interfacing with it. And urging this along is the corollary of *monozukuri*: *hitozukuri*, or 'making people'.

Traditionally, *monozukuri* for the individual artisan has been a state of mind focused on making things. The self-motivated individual has never departed from the *monozukuri* scene. However, it does not live or die with him either. It infiltrates the firm and the environment; its creed generates core competence. It is the essential element of supply and demand networking with a multitude of LMEs and SMEs populating the foothills. But its character is also affected by structural change and advancing technology. Demands for greater accuracy and precision make it difficult for the tiny workshop to survive because of the costs involved. Increasingly it is LMEs with the superior skills and management who are taking over as partners with the large assemblers. This also spells convergence of administration and LME interests. LMEs look to the administration for national *hitozukuri* programs; fusion of *monozukuri* and information technology going beyond the LME's chamber are essential. In fact, Japan's *monozukuri* and *hitozukuri* are driving its globalization impetus in which people and technology comprise a vast clustering on a national scale. This forms the backdrop for LME globalizing. The *monozukuri* man Horiba Masao has experienced all the stages.

11
Globalization and the Specialist

Globalization

Over the past two decades or so 'globalization' has acquired a hegemonic status as the all-embracing formulation of what is now evolving on a planetary scale. It simultaneously embodies the grandeur of the scope of undertaking and the compression of time and space which modernity has wrought, while also suggesting representations of how such a world is to be organized and sustained. The message it imparts is that economic, social, cultural and political change can no longer be understood as isolated phenomena. Moreover, these domains are increasingly less constrained by territorial and jurisdictional barriers (Olds *et al*, 1999). In other words, hitherto conventional terms of reference, geographical boundaries, and legal frameworks are being contested. A world regime which has hitherto been predicated on more or less sovereign states interconnecting with each other as their individual strengths and weaknesses dictate is now conceding ground to a diffused global order where the rules of governance are as yet in their early stages of germination. However, this evolution of circumstances, notably the compression of time and space, is to the advantage of the equipped specialist manufacturer. Globalization and the specialist are natural allies.

Mobilizing forces

The origins of modern globalization are to be found in trade liberalization, the revolution in information and communications technologies, and the spread of foreign direct investment. With respect to trade liberalization, following the end of the Second World War, GATT enabled the lowering of customs duties so that in the case of industrialized countries they fell on average from nearly 40 percent to under five percent by the end of the century, by which time the volume of international trade was 16 times the level it had been in 1950 (Nonjon, 1999). By comparison with the earlier relatively liberalized regime in the 19th century, the big difference now is the preponderant role deemed to be played by the large multinationals and the pressure they can exert on nation-states to assent to their demands, notably for accelerated deregulation, in line with their brand of globalizing impulses. These are globalized firms which manage on a worldwide basis the conception, production and distribution of their goods and services. Their mobilizing energy is devoted to the creation of a unified market with the global economy transformed into a single production and trading zone, monitored by a planetary regulatory and institutional framework in a world without borders. However, while this depiction is valid as far as it goes, it remains superficial. The specialist manufacturer offers another more detailed and localized dimension to the interpretation of this planetary phenomenon which is taking place at multiple levels.

The information revolution took incipient form in the early 1960s with the initial tentative steps in the convergence of related technologies. By the 1970s and 1980s this had evolved into the combining of computers and telecommunications into integrated systems for information processing and exchange (Dicken, 1992). The early 1960s was a time of internationalization when export flows were being developed. Emerging out of this activity was what in hindsight can be viewed as an intermediary stage of transnationalization which was more concerned with investment flows and establishing affiliates overseas. Then came globalization driven by the revolution in information technology which transformed both production and organization. If transnationalization denoted terrestrial reach, globalization bespeaks cyberspace reach. Coordination of production on a global scale through intra-industry and intra-firm linkages had arrived, with the IT industry itself as represented by the personal computer offering a perfect illustration: the CPU made in the United States, the memory in South Korea, the hard-disk drive in Malaysia, the LCD in Taiwan, and so on to the keyboard and mouse in China.

And that personal computer and the modularized components of which it is comprised also point to another mobilizing force of globalization: foreign direct investment. As implied above, FDI was already on the move in the transnationalization stage. This was when competition among the large assemblers of North America, Europe and Japan was intensifying, forcing them to shift manufacturing functions abroad on a massive scale. This exercise, with the subsequent injection of increasingly sophisticated IT inputs and networking agility, would eventually yield the type of organization epitomized by personal computer production. FDI *per se* has been defined by the IMF as 'investment that is made to acquire a lasting interest in an enterprise operating in an economy other than that of the investor, the investor's purpose being to have an effective voice in the management of the enterprise' (Hollerman, 1996:56). That is, FDI entails a firm from one country either buying an influential and possibly controlling interest in a firm in another country or setting up a subsidiary there on its own or jointly with other parties for the purpose of engaging in international production.

Foreign direct investment has seen explosive growth since the 1970s. During the final quarter of the 20th century it snowballed 25 times, its proportion to global GNP in terms of inward and outward stocks waxing twice as fast as imports and exports. Moreover, the momentum has not faltered. True, there was a downturn as of 2001, but by 2004 this was reversed with inflows estimated to have increased by six percent and then boisterously by 29 percent to US$916 billion for 2005 (UNCTAD, 2005 and 2006). FDI has been adopted as a highly effective value-adding process. Over recent times, moreover, it has been joined by other increasingly elaborate value-adding devices entailing intermediate production and cross-border specialization through such instruments as strategic alliances tailored for specific purposes, as again illustrated by the personal computer example above.

Dunning's eclectic paradigm theorizes on the motives and logic with respect to the individual firm thus. It possesses *ownership advantages* in the form of, say, technologies and managerial skills, which enable it to countenance manufacturing abroad. It identifies certain *location advantages* associated with a given country, such as the availability of lower-cost manpower or natural resources, which merit setting up there. At the same time, the firm elects to benefit from its *internalization advantages* by creating an internal market through the act of directly investing itself in that country in order to control key sources of competitiveness while protecting its proprietary knowhow and technologies (Dunning, 1980). The specialist manufacturer of core components, then, sets up in a country which is promoting its cheaper factors of production to assemble televisions and motorbikes, say. In so doing the manufacturer enhances the host country's prospects while protecting its own interests. In one sense this could imply that the specialist is being captured by globalization in that it cannot afford to stay outside the loop. But this is to overlook the other mobilizing forces of globalization: the speed of transportation and the growing sophistication of communications. This is why, depending on how they interpret their options, some specialists go for a physical presence worldwide, others concentrate heavily in key global regions, and yet others conduct all their business from home base. Globalization has shuffled the options for the specialist manufacturer.

Implications for the Japanese LME

In the revised edition of his book on LMEs brought out in 1990, Nakamura Hideichiro stated categorically at one point: 'Whether or not they firmly establish themselves as international specialist manufacturers will no doubt be one of the decisive factors for the development of LMEs from here on'(Nakamura, 1990:353). The postwar version of internationalization had, of course, been in the works for a number of decades by then. As we have seen, Pentel was already hitting the export trail by the middle of the 1950s and many more were to follow in the 1960s. But whereas the government encouraged exports it only finally conceded to capital liberalization – and therefore to outward FDI – to comply with IMF and OECD stipulations, having crawled through four stages of relaxation of restrictions from October 1969 to July 1972. The eventual outcome was an about-turn in faith. Having been a minuscule FDI participant in the 1960s and still insignificant in the mid-1970s, in terms of annual flows of outward investment Japan was third in the world in 1985 and number one in 1988. By the 1970s the need to realign comparative advantages which would inevitably entail the cultivation of an international division of labor was becoming increasingly self-evident anyway. As one commentator observed at the time, for Japan 'overseas production has suddenly emerged as a national requirement encompassing practically the entire spectrum of her industries and enterprises, small and large alike' (Ozawa, 1979:228).

Japan adjusted, then, to external pressures and ongoing necessity. But although the forces of worldwide economic integration have since then become progressively more insinuating, Japan has not swallowed the package wholesale. It has resisted to a considerable extent, for example, the push towards financial dominance in corporate practice championed by the United States and Britain. Come to that, leaving aside foreign imports Carlos Ghosn at Nissan and Howard Stringer at Sony, neither has Japan been much taken by the universal executive

American-style. It remains product-oriented and particularistic. Moreover, Japan has promoted its own prescriptions for internationalization as it has subsequently morphed into globalization. It has, for instance, made no bones about assisting its companies in their overseas ventures through such bodies as the Export-Import Bank, providing them with loans, and Japan External Trade Organization (JETRO), assisting them to set up in foreign locations. And over more recent times, the samurai authorities have been keen advocates of what they see as Japan's strengths epitomized in their version of *monozukuri*, not least because embodied in these strengths are the potential regarding Japan's own particular interpretation and realization of globalization. In a word, they are conscious of the desirability of sustaining as far as is justifiable the distinctive institutions, policies and norms of the country and indeed promoting them in a wider world where feasible. One of these institutions is the Japanese LME.

And the *Japanese* LME emanates, by definition, from a national context. This national context can be likened to a runner in mid-flight along the track. The runner's build, style and performance are what distinguish him from the other runners. One distinguishing feature of Japan's national context is the LME and how it has been derived within that context. Deriving from a singular historically and socially mediated environment, LMEs had become key players in the matrix of relationships and networks that had evolved as Japan's economy developed. What is more they were to reinforce their relevance in the decades following.

However, it was a relevance in an increasingly international setting. Contrasting two snapshots, one taken in the mid-1960s and one taken around the middle of the current decade, can illustrate this. By the mid-1960s the LME were ensconced within a thoroughly national context. Whether the producer of intermediary items or end-user goods it existed within the realm of a tiered, inter-firm division of labor and all the business relations incidental to it. The LME coordinated with others flexibly and more often than not on the basis of trust rather than written contractual agreements. This notion of trust, however, has to be qualified. After all, legislation had to be passed in the 1950s to prevent large corporations delaying payments to smaller suppliers and squeezing their margins. Actually, to enjoy the benefits of trust in Japan the company had to win its spurs, and becoming an LME, as many firms did in the 1960s, was equivalent to that. So trust, in fact, was based on reliability, predictability and the Japanese aversion to the unexpected. Having gained access to a network of trust the firm became quasi-indispensable and its position was virtually assured. It was part of an inviolate national structure viewing the outside world – if required – through a Perspex window.

Globalization has diluted the national context on which this arrangement was founded, as our second snapshot reveals. The LME, along with corporate Japan in general, has long since overcome its reluctance to invest in and manufacture abroad, although this approach is still regarded as an option rather than an absolute necessity. Its attitude is perhaps best expressed by saying that it is far less conscious of looking through an interface represented by the Perspex window. Much has happened to induce this change. The organization of domestic production has undergone quite a dramatic upheaval. This is not just a matter of the large corporation, for example, reducing and restructuring its domestic production; it has also led to the transformation of production linkages right down to the subcontracting small firm. Caught in this vortex the LME may have had to set up manufacturing abroad in compliance with client wishes, diversify its customer base especially overseas, or resort to cheaper foreign parts suppliers to sustain its competitive edge. The tentacles of division of labor are increasingly borderless; the boundaries of trust have been tested and rearranged. What is more, alien business practices – actually Anglo-Saxon – are ever more persistently knocking at the door, letting it be known that the company is to be run mainly, if not exclusively, for the benefit of the shareholder over the interests of the employee, the manager, and the customer, in stark contradistinction to what many an LME thinks it exists for.

And yet, there are other things to see in that snapshot of circa 2005. For one, there is ample evidence to suggest that Nakamura's prescription for the survival of LMEs has reaped returns: there are many Japanese LMEs which have become international specialist manufacturers. Moreover, much of the background to this is to be found in the national context. In the first snapshot the nation itself can be seen as the chamber, in this case an incubating chamber in which LMEs could grow. But what must be remembered is that within this chamber competition was fierce; many smaller firms fell by the wayside and bankruptcies were rife. Here was the rivalry which coexists in Japan at an often high level of tension with collaboration, not unusually deliberately cultivated also within the company's chamber to pit individuals and departments against each other even as they work together in the interests of a common goal.

Rivalry and collaboration spurred innovation as LMEs looked to elaborate their core competences, as well as to enhance integration of expertise through networking and geographical concentrations like Higashi-Osaka and Tokyo's Ota Ward. Not that the international was ignored; far from it. To cite two firms already encountered, Shinagawa Refractories and Akebono Brake were soon in the market for foreign technologies in the 1950s, while the ideas and practices of quality control were easily recognized and eagerly adopted as natural concomitants of *monozukuri*. But – and this is the point – *monozukuri* it was and *monozukuri* it remains, as is clear in the second snapshot. Even as they achieved greater autonomy and more readily recognized corporate identities with their proprietary products and enlarged customer bases, and even as they progressively ventured abroad from the 1970s on, most Japanese LMEs have retained their Japaneseness, often indeed operating overseas still to a considerable extent within their national context. Compromises they have had to make to be sure, but they still come armed with their nation's singular business practices and attitudes predicated on *monozukuri* and do not defer outright to some presumed overriding logic of globalization. Or, put another way, they present another face of globalization in their own image. The national context may be diluted but it is far from expunged.

The Japanese LME ...

So there are two outstanding features of the Japanese LME as it moved into the 1970s: having been, first, born of a Japanese context it was, second, now progressively achieving greater autonomy. In looking at this LME as it subsequently internationalizes and globalizes the aspects of these features can be ticked off, starting with the national context. In fact, the national context itself began to internationalize, taking with it its singular culture and mentality. Moreover, firms and institutions, being attached to it, did not venture out in total isolation. Rather, they collaborated with fellow nationals to steady their footing in the unknown. This has been particularly salient for firms coming from the society which is Japan, at once inordinately insulated from the outside world while enjoying a high level of trust as defined here within its own boundaries. What this meant was that, while like virtually everybody else, the Japanese were wary of foreigners on initial contact, in contrast with other people, notably Americans, they were more likely to trust their own. Japanese firms coordinated and emulated; the herd instinct was strong. They did many of the same things for the same reasons.

A panoply of guidance and assistance is available for the exporting and overseas-investing Japanese firm, emanating from both the public and private sectors, and not infrequently coordinated between the two. The Export-Import Bank of Japan and Overseas Economic Cooperation Fund (OECF) provide finance, METI insures against losses, and the government uses foreign aid in the form of technical training and infrastructural development to stimulate FDI (Arase, 1995). JETRO runs incubation centers in host countries to assist Japanese firms setting up there, and together with the Japan Overseas Development Corporation (JODC) and the Institute of Developing Countries, collects and disseminates economic and political information about countries worldwide. Apart from the banking sector *per se*, the most substantial private-sector contribution has been afforded by the ubiquitous general trading companies, or *sogo shosha*, which, among other things, provide financial services in the form of loans and venture capital, information on production markets, risk-hedging facilities, assistance in organizing contracts and linking up with suppliers and customers, and office processing of paperwork, insurance and transactions (Young, 1979). Together with the banking fraternity, the general trading company is in effect the samurai's private sector linchpin, combining Japan's national interests and its own on the international scene. This is symbolized – to give a very recent example – by Mitsui's recent talks with Russia's state-owned operator to launch a trans-Siberian freight service clipping 40 percent off the time taken by sea to Moscow and St Peterburg, aimed at automakers and other Japanese companies advancing into Russia (Nikkei, 2007c). It is no longer absolutely inevitable that this Japanese clientele will simply fall into line, but the odds are high, and this reflects the strong inclination of Japanese business to work closely together on foreign soil as they do back home.

.... the autonomous LME

In a general sense, Japanese LMEs have not deviated from the pattern as outlined in the conventional tale of their country's corporate globalization, nor have they foresworn the accommodating framework of home-country institutions presented to them. However, although 'Japan, Inc.' certainly does play its part, this is far from being the whole story. The autonomous LME also pitches in. Here was destiny in the making. And the reasons why Japanese LMEs increasingly embarked on a more independent course of action as they internationalized especially from the 1970s on can be found not only in these specialist enterprises themselves, but also in the situation created within their home country, coupled with the transformation just beginning to sweep across the world already alluded to.

By the beginning of the 1970s, many LMEs were at the point of achieving a status of significance within a national economy which had just two or three years before also achieved the status of second largest economy in the free world. In stylized terms these LMEs had honed their core competence expertise to the point that they could satisfy the demanding standards of this aggressively advancing economy both for intermediate and finished goods. And the fact that they could meet such standards meant that they could be competitive in most if not all overseas markets which could accommodate the high quality of their output. Even if their international experience was still tentative their potential was no longer simply parochial. As such, with perhaps a little help from the national promotion organizations mentioned above, they could pinpoint with a fair degree of accuracy where they wanted to be. Limited in size and scope though they may have been, the superiority of their output ensured them a hearing among the elite of manufacturing as well as in less advanced but emerging economies which valued and needed their expertise. This was the pull side predicated on what they offered. It had already found expression in exports and was now looking increasingly to FDI.

But there was also – and this is the second reason – a push side emanating from what Japan had become which was especially pertinent as far as FDI was concerned. Mounting manufacturing costs, notably with respect to pay checks, were of course one factor, but there were also more complex economic and social mechanisms at work. Imminent saturation was one, although this is not merely to be interpreted as a surfeit of production. It has as much to do with realigning rigidities within the structure as a whole. This realigning during the early 1970s took the form of intensifying cross-shareholding among companies seeking protection from take-over by large foreign predators as capital liberalization supposedly opened up the Japanese economy. The gentleman's agreement was that the cross-holdings of shares would not be traded without due notice, thereby mitigating the chance of undesirable outsiders buying them up. It also encouraged an inclusive business coziness, which limited the opportunities for up-and-coming suppliers to break in. Inclusiveness excluded. Time and added expertise were to open the door of exclusion to the specialist over the coming decades, but there was another rigidifying element in play as well.

This was hierarchy, both economic and social. In this book the Japanese inclination to hierarchy has been polarized into simply samurai and artisan, but in fact it is far more intricate. There has been some unraveling of the system since the 1960s and 1970s, but at that time it was all but inconceivable for a graduate of a lowly technical college, unlike a graduate from Tokyo University, to become president of Mitsui; it was almost unheard of for a Mr. Nagamori, still in his 20s, to approach the equivalent of IBM with the innovations of his pipsqueak outfit called Nidec stuck among the paddy fields and expect to be taken seriously. Companies had to know their place and go through the accepted channels. The position of the company was enshrined in the sanctity of market share. In fact, to dwell too much on government support packages for smaller firms is to some extent to miss the point as to the restricted entry in an economy dictated by stratification. Such support in this light can be interpreted as compensation for not disturbing the order. True enough, Sony and Kyocera made it to the top ranks (as Nidec is doing now) and Honda simply bucked the system when it added automobiles to motorbikes in the 1960s strictly against the wishes of the administration. However, what happened to many Japanese LMEs from the 1960s into the 1970s and thereafter is that they grew *with* a growing economy, but

Figure 11.1   FDI Push and Pull Factors for Japanese LMEs

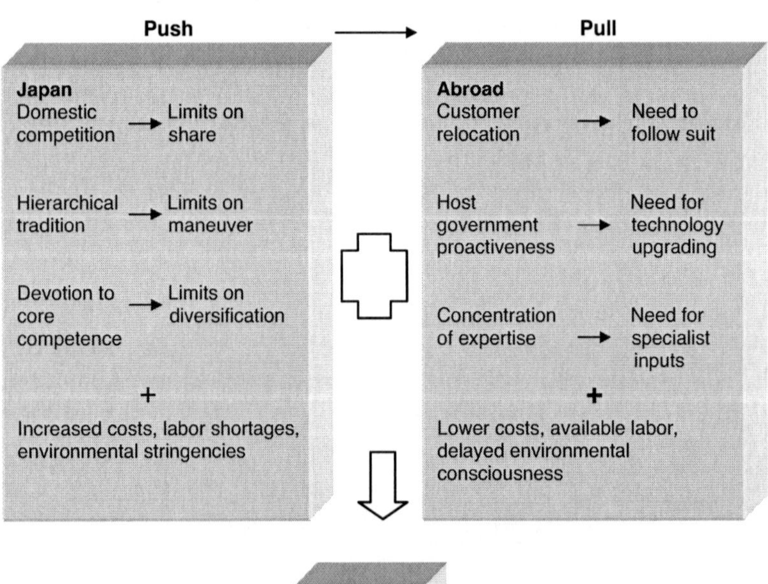

with their ranking in the whole stratification *within Japan* budging hardly at all once they had attained LME status.

With the economy becoming increasingly sophisticated, horizontal diversification to take on evolving potential as Mabuchi accomplished with its DC motors, was one solution for the undaunted, ambitious LME. But there were fewer options for Fujitec. Having been established in 1948 by Uchiyama Shotaro for the research and development, manufacturing, sales, installation and maintenance of elevators, Fujitec rapidly moved into the LME ranks, establishing sales offices in all major cities in Japan by 1950 no less, but it was nevertheless in what is essentially a large corporation industry. It could have challenged the competition head on, but this society of trust did not take kindly to the violation of market share. It could diversify along its core competence theme, which it did with escalators and multistory car parks, but its rivals were still big. No doubt this induced Mr. Uchiyama's prescience, for as early as 1964 he is quoted as saying: 'The world is just a single market. In today's open economy, I believe the key to our great success is to embark on an aggressive strategy to take business to the global market' (Fujitec, 1998). And it was that year that Fujitec set up its first overseas subsidiary in Hong Kong.

In this process, what Fujitec is illustrating in a particularly stark fashion is what these push and pull reasons have so far left out: the positive, motivating force to internationalize inherent in the LME based on the intensity of its core competence which results in its outgrowing the potential of its original home market. This, as already noted, is one of the distinguishing characteristics of the universal LME. Many Japanese LMEs were primed for this by around 1970, and were at the point of being stimulated by inviting global horizons. So they started exporting – if they had not started already – and investing in earnest. At that stage, their exports went mainly to North America and Western Europe and their direct investments for manufacturing were heavily skewed to the United States and the Asian NIEs – Taiwan, Hong Kong, and Singapore especially. The United States investments were almost exclusively aimed at market proximity both for intermediary and final goods, while those in Asia could be to take advantage of lower production costs for exporting to the advanced countries, but were also made to service Japanese clients who had moved there, to provide nascent local industries with advanced components, materials and equipment, and to satisfy new end-user needs as they began to appear.

Box 11.1   Shimano starting out in the world

The back page of the financial daily Nihon Keizai Shinbun, or Nikkei, always features the personally related curriculum vitae of a high achiever in the form of a series lasting for around a month. July 2005 was the turn of Shimano Yoshizo, third son of one of the early LME founders, Shozaburo, encountered in Chapter 6. Yoshizo is now chairman of the family business, but when he was young he was the very initiator of its internationalization, homing in as it did first on the United States and then Europe, as a second market, and East Asia, as a manufacturing base. Subsequently a rendition of this serialized life history was published in English as a book entitled *This is My Road*. Drawing from both we can get a glimpse into how Shimano started to globalize (book page numbers in parenthesis).

Getting going

Japanese bicycle exports to the United States began to pick up from the early 1960s. 'Some of the bicycles exported to the U.S. were equipped with our company's 3-speed internal hub gears, but it was difficult to export bicycle parts alone (53) …… When Shimano participated in the International Toy & Cycle Fair in New York as an exhibitor in 1961, our exhibits attracted little attention. In those days, most Americans still thought lightly of Japanese products, considering them utter junk and cheap goods of poor quality. Before long, though, Shimano's 3-speed hub gears, which were compact and performed well, began to attract interest from complete-bicycle manufacturers in the U.S. (53–4)' …

There were eight major bicycle manufacturers in the U.S. Shimano commenced negotiations with Columbia. 'Shozo, then Shimano's president [and Yoshizo's eldest brother], never compromised on price, because he thought that Shimano should act consistently as a leading manufacturer in developing our business overseas (55).' In the summer of 1963, Columbia's president visited Japan to conduct final negotiations. 'In the final stage of the negotiations, Shozo proposed a one-year contract with advance payment. At last, after some heated exchanges, the two companies concluded an agreement for 100,000 units a year (55).'

Shimano was on track for the American market, but still based on the principle laid down by founder-father Shozaburo: 'Produce it by yourself. Sell it by yourself.' And the person to sort that market out was to be Yoshizo, himself. His brothers reckoned it would take him three years to get things going. So on 1st July, 1965 Shimano America Corporation (SAC) was set up staffed by just three people, including Yoshizo as president.

Box 11.1   Shimano starting out in the world – *continued*

Doing the rounds

The American manufacturers were scattered all over. 'Some manufacturers along the East Coast were within a four- to five-hour drive from the SAC office .... I visited many bicycle manufacturers carrying a projector and screen for presentation purposes, as well as samples (59).' They accepted the product and thought the price was right. But they wouldn't buy. The reason they gave was that they had no customer who wanted a bicycle fitted with Shimano parts. The manufacturers knew about Shimano alright, but it had no name recognition among the retail outlets and consumers. It was up to Shimano to create the demand first, as one manufacturer bluntly put it.

This provoked the response. 'I thought it was vital to ensure that bicycle retailers knew and understood Shimano products. I visited buyers at mass merchandisers and held seminars for local bicycle dealers. My bag was filled with various items including product brochures, sample parts and design drawings.'

'At the clinics, I explained to retailers the concept under which Shimano products had been manufactured, as well as describing the materials and manufacturing methods used for them. I explained that users could use our products with the fullest confidence, as our products were of high quality and spare parts were also available. I disassembled and reassembled our 3-speed internal hub gear in front of clinic participants (60–1).'

The breakthrough came with Sears, Roebuck, then America's largest retailer. It had a state-of-the-art test laboratory which it used for testing its own brands. The company liked Shimano's product, and started to use the 3-speed internal hub gear for its children's models, children accounting for 70 to 80 percent of the U.S. bicycle market at the time. This heralded the start of Shimano's business dealings with most of America's bicycle manufacturers.

Things were buzzing and the New York office was in contact with headquarters in Sakai just outside Osaka every day. 'We had no copy machine or fax and telephone charges were so high we did not make calls except in cases of urgent necessity. We sent telegraphs using abbreviations that we devised so that we could save on the number of letters and thus reduce cost.'

Box 11.1   Shimano starting out in the world – *continued*

'Whenever, on the basis of our U.S. customers' requirements, we requested trial manufacturing and improvements in our products, the headquarters responded promptly and sent us workable samples as soon as possible. In due course, when I visited bicycle manufacturers and retailers, they began to say, with a look of expectancy, "What did you bring with you today?" I took various things out of my bag, including unique new products, improved parts and upgraded design drawings, and showed them to manufacturers and retailers. They were sincerely impressed with these things, and surprised by our prompt response to their requests. My heavy bag became known as the "magic bag" (62–3).'

By the 1970s, a new market was developing: adults. Having hitherto been regarded as a children's toy in the United States, the bicycle was becoming the vehicle for the fitness and environmentally conscious of all ages. New markets called for new research. 'I wanted to know about trends in retailing and the entire bicycle industry; however, in the U.S. there were about 6,000 bicycle dealers, which were scattered across that vast country. As I thought it was impossible for SAC members alone to visit all of them, I asked the headquarters in Japan to select younger staff, mainly in their late 20s. I formed sales promotion teams, called "caravans", to make the rounds of bicycle dealers throughout the U.S.'

'"Caravans" composed of two members traveled across the United States on a rotating basis for periods of six months. Their objective was not to sell products, but to provide after-sales service, deal with complaints, introduce our products, give hands-on help repairing our products and collect information.'

'In January 1971, the first team departed on its trip around the U.S. in a 4,000-cc station wagon loaded with various items on its rear seat, including our newly-introduced products, various components, and repair and service tools. .... Upon arriving in a town, the caravan team would consult the telephone directory and visit the bicycle dealers listed in the directory (69–70).'

Big orders

In America by the 1970s what had come to be required of an adult's sports bike was that it be lighter and faster. There was also a shift in preference from the internal to external type, and from 3-speed hub gears to 10-speed derailleurs. The latter had originally been developed in Europe for racing and had since then come into general use. When

Box 11.1   Shimano starting out in the world – *continued*
they started to become fashionable in the United States, the European manufacturers had a head start and took the lead. But it seems they mistook their strategy for America, partly because they had their hands full coping with the European market. However, they did not anticipate the great boom that was about to hit America. But Shimano was different.

'Reports sent from our caravan teams confirmed the trend toward 10-speed bicycles even in small towns in states such as Alabama and New Mexico. I firmly believed that it would be a good opportunity for us. I informed the headquarters in Japan about this and asked them to tool up for increased production (70).'

Just then, in 1970, Shimano had completed its Shimonoseki Factory in Yamaguchi Prefecture at a cost of five billion yen. The reasons for this huge investment were, first, that the company's sales in the United States were increasing and, second, Europe was coming on steam as the next market to tackle. In fact, Shimano's sales of its 10-speed derailleurs exploded in the United States, peaking in 1972 and 1973. Production could not keep up with demand. 'Once, we even chartered a cargo plane to carry products to the U.S. European bicycle parts makers could not respond to the U.S. demand like we did (77).'

FDI

When the Shimonoseki plant was built, land prices were starting to rise and it was becoming difficult to hire people. Once the idea had taken hold that the next step was setting up overseas, things soon started moving.

'My second-oldest brother, Keizo, recommended Singapore. I went there with him and visited the government department in charge several times. As a condition for approving our factory construction plan, the person in charge proposed that Shimano establish a joint venture with a local company.'

'We had no objection to this proposal. However, since we anticipated difficulty in our first attempt at overseas production, we explained that our planned factory's prospects for instant success were poor. Therefore we requested that the department find a partner who could bear with us even if the planned factory was continuously in the red for several years (78–9).' The upshot was that the authorities decided to allow Shimano to set up a wholly-owned subsidiary and the plant started operating in 1974.

> **Box 11.1   Shimano starting out in the world** – *continued*
>
> Footnote: Today, in addition to Singapore, Shimano has overseas manufacturing facilities in South Korea, China, Malaysia, Indonesia, the Czech Republic, and the United States; it has additional sales subsidiaries (for bicycle components and fishing gear) in Sweden, the United Kingdom, the Netherlands, Belgium, France, Germany, Italy, Finland, Russia, and Canada.
>
> Shimano (2005), Shimano (2008)

Summary

Globalization is a blending of economic, social, cultural, and political changes which can no longer be taken as isolated phenomena. It has been mobilized by increased trade in the postwar era, the information revolution which can be traced back to the early 1960s, and FDI. The latter has seen explosive growth since the 1970s, up 25 times during the last quarter of the 20th century. The specialist manufacturer was well primed to take full advantage of the globalizing environment presaged on these mobilizing forces.

With liberalization at the beginning of the 1970s Japan fast embraced FDI, actually becoming number one in 1988, as the need for international division of labor became increasingly self-evident. But it came up with its own version of internationalization. It was not convinced, for example, by Anglo-Saxon financialism or the universal executive. It has unabashedly assisted its firms to set up abroad and looks to promote its own institutions where possible. One of these is the LME. Having emanated from a distinctive national history and culture, however, the LME is now posited in an international setting in which globalization has diluted the national context where the original arrangement took shape. New production formats are called for and alien business practices cannot be ignored. On the other hand, the LME have reaped returns predicated on the national atmosphere of rivalry and collaboration it has grown up in. Upon going abroad LMEs have taken their *monozukuri* along with them as an inevitable part of their national identity. They have often acted in concert and have had at their disposal a panoply of guidance and assistance provided by government and quasi-government organizations, banks, and general trading companies.

That having been said, the LME has also acted autonomously urged on by both pull and push factors. On the pull side have been the high quality of its products and expertise which have attracted the attention of foreign markets. On the push side have been rising domestic costs, imminent saturation of the home market, protective cross-shareholding precluding newcomers, rigidly sustained market shares, and fixed status. Due to their intensive concentration on their core competences, many LMEs were beginning to outgrow the purely national market by the 1970s, and like the universal LMEs encountered earlier were primed for internationalization. Shimano provides an example of how this was done.

12
A Population of Globalizing Japanese LMEs

Timing

This chapter is based initially on data available for fiscal 2004. It was around this time that Japan was hitting a sort of peak in the Japanese way of doing things. For a start, the prolonged recession going back to the advent of the 1990s had been buried and the country had recovered its economic dynamism and sense of purpose. Part of the motivation for this was a reassessment of what Japan stood for in itself and within the world economy. First and foremost, it had been concluded, Japan was the land of *monozukuri*, the land of high-quality production. This refurbished self-differentiation in turn was inducing a reinterpretation of what globalization meant for Japan and how it could be handled. Gone was the feverish exit in the quest of cheap factors of production. Less acute were the fears of the alien prescription of a forced march overseas. Increasingly greenfield plant investment was as likely to be back in Japan as abroad. Advanced automation and communications coupled with a renewed appreciation of the advantages inherent in the high levels of integration of specialized suppliers and well-trained personnel encouraged this inclination.

In other words, from this standpoint it would appear that globalization had become more manageable and easier to orchestrate from a home base which, far from hollowing out, was reasserting itself as an active contributor at the apex of an elaborated international division of labor. However, globalization was not born to stand still. It was also around 2004 that fresh evidence was emerging that, despite a newfound self-confidence in its own potential, Japan was ever more obliged to take note of and act upon what was happening outside and what was encroaching inside. Happening outside were fast-developing economies which, in addition to China, now included Brazil, Russia and India. These BRICs so-called had to be embraced if Japan was to retain its standing. Encroaching inside were new business practices and attitudes, like the demands from a growing army of foreign investors for enhanced dividends. These represent incipient challenges for the population of 429 LMEs introduced here.

In this chapter we start by looking at what this population had achieved as globalizers by the peak year of sorts of 2004. But, again, it is as well to have a good notion of what kind of firm we are talking about, and we do this by comparing the 429 LMEs as a population with Toyota Motor Corporation. It is apparent that the sample population as a whole has been concerned to establish its presence in the advanced world as represented by North America and Western Europe and this is in keeping with the approach of universal LMEs in general. However, it differs from most Western LMEs in the degree of attention, especially for manufacturing, that it has paid to the region of which Japan is a member: East Asia. Moreover, Japanese LMEs tend to be numerous and highly effective in certain industries. Machine tool manufacturing comprises one such industry which has put its own distinctive stamp on globalization.

A global corporation: Toyota

Toyota Motor Corporation is currently the 'jewel in the crown' of Japanese industry. In terms of revenues it is sixth in the world and first in Asia (Fortune, 2007). At the same time, unlike the comprehensive machinery or electronics manufacturing giant, everybody knows exactly what it produces. The family presence is still strong, with a grandson of the founder in line to be a future president. So it is in a sense like some vast LME, concentrating unequivocally on its dominant core competence: the automobile. Moreover, there are other similarities to the stylized LME. For one, Toyota is a professed devotee of *monozukuri*, based on 'lean production' which combines the best of craftsmanship and mass production (Womack *et al*, 1990). There again, its current president, Watanabe Katsuaki, is a renowned stickler for cost cutting and is possessed of an overriding concern that Toyota does not degenerate into complacency and contract 'the big corporation disease'. This is a remote possibility for the foreseeable future. In fact, Toyota makes no secret of its goal to overtake General Motors and become the world leader, something it is very close to accomplishing having recently recorded a year-on-year increase in profits of 32 percent and larger sales than its great rival for that quarter (Nikkei, 2007d).

429 Japanese LMEs

The LMEs for this analysis qualify for selection on four grounds. First, they feature in two publications issued in 2004, these being the English-language Japan Company Handbook (JCH), which details all companies on the domestic stock exchanges, or *Mijojo Kaisha Ban* (MKB), a compilation of information covering some 7,000 unlisted firms, plus *Kaigai Shinshutsu Kigyo Soran* (KSK), which is a roundup of data concerning Japanese company overseas affiliates and representative offices. In other words they appear in either JCH plus KSK or in MKB plus KSK. Second, they have a *domestic* workforce as stated by these publications of between 200 and 2,000. The number of employees within Japan is used because it has been the most consistent defining feature of these companies over a prolonged period. Capital and sales have increased in step with the growth of the national and, increasingly, international economy, while home-based staff have remained relatively stable by comparison. As already noted, rather than internalizing, not a few LMEs have realized business expansion by close collaboration, stock acquisition, and the creation of subsidiaries while maintaining the core company as a distinct and separate corporation. It is these core companies with their core competences that we are looking at; Nidec Corporation, but not Nidec Copal, Nidec-Read, or Nidec Tosok included or in addition. Developing overseas operations has been the other means of expansion and this has led to some striking corporate configurations: in 2004, labor-intensive DC motor maker Mabuchi had just 804 employees in Japan although its wage bill extended to over 40,000 persons, most of them in China. It remains an LME by this definition. Third, the company is not a subsidiary nor is it closely integrated into any corporate group other than what it creates itself, which means that Nippon Kokushuha Steel as part of the Kobe Steel group and 51 percent owned by that corporation does not qualify. Fourth, no entity, other than founder- or company-related, has an ownership share of 20 percent or more of the company, which means that Akebono Brake, albeit 14.2 percent owned by Toyota, does qualify.

Size and shares

Although it is difficult to use specific capital stock amounts over a period of time to define LMEs, capital at a given time can still be cited along with employment figures for contrasting the LME population with Toyota, for example, as illustrated in Table 12.1. Clearly the average LME in terms of both capital stocks and workforce is a mere fraction of the size of this large corporation, being less than two percent in both

Table 12.1   LMEs vs. Toyota: Capital and Employment (2004)

		Capital (million yen)	No. of Employees
429 LMEs	Total	3,272,701	335,637
	Average	7,629	782
Toyota Motor		397,050	66,099

Sources: Japan Company Handbook, Summer 2004; MKB (*Mijojo Kigyo Ban*), Autumn/Winter 2004.

Table 12.2   LME Typology

	Non-Family			Family		
	LMEs	Av. Capital	Av. Emp.	LMEs	Av. Capital	Av. Emp.
JCH(1) – 1st Section	129	13,003	1,085	119	9,812	857
JCH(2) – 2nd Section	23	10,096	439	98	3,105	502
MKB – Unlisted	12	1,633	592	48	740	574

NB: Average capital in million yen.
Sources: As for Figure 13.1.

cases. Also to be noted is how small, in fact, the total LME workforce is given the key functions and diversity of production LMEs perform within the national economy and beyond, evidencing the intense concentration of capability as that economy has progressed. This factor alone strikingly underlines the concentration of specialized expertise that these LMEs represent. The data is then broken down in Table 12.2 to produce a typology based on listed and unlisted LMEs on the one hand and family and non-family LMEs on the other. In addition, listed companies are divided into LMEs in the First and Second Sections of the national stock exchange system.*

A family LME is defined here as one in which the founder and/or founder-family still display an active involvement in the company in their roles as chairman, president or leading shareholder. The reason for highlighting founder and family is, needless to say, to reemphasize the important part they have played and still do. In fact, quite a few of

---

*Listed LMEs here may be registered with the Tokyo Stock Exchange, which is by far the largest, the exchanges of Osaka, Nagoya or Fukuoka, or the Japan Association of Securities Dealers Quotations (JASDAQ). The stock exchanges have first and second sections and JASDAQ is included in the second section here.

the non-family LMEs bear the name of their founder, as in the general machinery category, the Amada of Amada Isao, the Okuma of Okuma Eiichi, and the Takuma of Takuma Tsunekichi, for instance, but the family name no longer features among the officers and owners as indicated here in these particular cases. What the non-family LMEs in the First Section demonstrate, however, is the evolution of the LME along the lines prescribed by Nakamura. They suggest that as the cleavage between ownership and management has widened these companies have grown in terms of capitalization, workforce, and status. The fact that the majority of non-family LMEs have been elevated to the First Section is further evidence for this.

Nevertheless, the persistence of family involvement is incontrovertibly demonstrated by the large number of family LMEs in the First Section. This in turn organically links up with the fairly substantial numbers of family LMEs in the Second Section and among the unlisted companies, some of whom have since 2004 joined the First Section ranks. Moreover, looking diagonally across from the First Section on the top left to the Second Section and unlisted companies on the middle and bottom right it is possible to discern the continuum in the Japanese economy from the samurai-orchestrated to the artisan-instigated, and how the individual LME is constituted as a consequence.

This can be illustrated by the composition of ownership and the role of financial institutions within that. Detailed information is not given for unlisted companies in MKB, but it is for the 369 listed companies in JCH. Among other information documented in JCH is a list of major shareholders at of the end of the most recent account settlement term – usually ten of them, their holdings being expressed as a percentage of the total. Hence, as noted above, Toyota is listed as a major shareholder of Akebono with 14.2 percent. Ownership patterns are apparent which are consistent in sequence from non-family First Section, family First Section, non-family Second Section, and finally to family Second Section. Hence, for example, the major shareholders own an average of 38.1 percent of the stock of non-family First Section LMEs and 51.1 percent of family Second Section LMEs, indicating that ownership is more dispersed the more established the company becomes in the national business hierarchy. Banks and trusts as shareholders reflect the same inclination. Calculations for 2004, for instance, show the Mizuho Financial Group had a 4.3 percent share in non-family First Section LMEs and 3.4 percent in family Second Section LMEs. A sure sign of establishment status over recent years, moreover, is for companies to be invested in by The Master Trust Bank of Japan (MTB) and Japan Trustee Services Bank (JTSB), both jointly founded by large banking and insurance institutions during the financial reconstruction extending into the new millennium. Over three quarters of non-family First Section LMEs are counted among this select group, whereas the figure is less than a quarter for family Second Section firms, for whom the share percentage of these trusts is considerably lower also.

The globalizing Japanese LME

Description

Table 12.3 summarizes the LME population in the context of the industries in which they are engaged. The industry categories are those

Table 12.3   429 Globalizing LMEs: A Summary

Industry Category	Family	Non-Family	Listed JCH(1)+JCH(2)=Total			Unlisted (MKB)	Totals
Food	12	6	13 +	5 = 18		0	18
Textiles and Apparel	6	14	16 +	2 = 18		2	20
Pulp and Paper	2	1	1 +	2 = 3		0	3
Chemicals	35	31	42 +	16 = 58		8	66
Pharmaceuticals	3	5	7 +	0 = 7		1	8
Oil and Coal Products	0	2	0 +	1 = 1		1	2
Rubber Products	9	3	4 +	3 = 7		5	12
Glass and Ceramics	5	5	8 +	2 = 10		0	10
Steel Products	2	4	5 +	0 = 5		1	6
Nonferrous Metals	5	4	3 +	2 = 5		4	9
Metal Products	9	7	6 +	6 = 12		4	16
General Machinery	57	33	52 +	24 = 76		14	90
Electrical Machinery	68	24	56 +	30 = 86		6	92
Transport Equipment	16	12	13 +	11 = 24		4	28
Precision Instruments	12	8	7 +	6 = 13		7	20
Other Products	24	5	15 +	11 = 26		3	29
	265	164	248 +	121 = 369		60	429
	429					429	

employed in the JCH and MKB publications. As can be seen, there are in all 265 family LMEs and 164 non-family. Alternatively expressed, there are 369 listed LMEs and 60 which are not listed. With reference to Figure 12.2, while the First Section listed companies are evenly spread between family (119) and non-family (129), in the Second Section the companies are predominantly family, accounting for 98 out of 121. The unlisted LMEs are also mainly family – 48 out of 60.

As for the industrial categories, two especially stand out: general machinery and electrical machinery. This is evidence that this kind of industry is more easily established and sustained by LME founders than, say, oil and coal, an observation further supported in this table by the fact that twice as many machinery manufacturers are family than non-family. The general machinery firms include Koike Sanso and the other machine tool makers we saw founded in the first half of the 20th century; the electrical machinery makers include Mabuchi and others of the class of the 1950s, extending to the later arrivals Nidec in the 1970s, and Allied Telesis in the 1980s. Turning to the other categories, the companies established by the six founders introduced in Chapter 6 afford a representative spread: S&B Foods for food products, General for pulp and paper, Tanaka Kikinzoku for nonferrous metals, Shofu for precision instruments, Shimano for transport equipment, and Mizuno for sports goods in the 'other products' category.

Numbers

Table 12.4 compares the 429 LMEs with Toyota with respect to overseas affiliates and representative offices according to the *Kaigai Shinshutsu Kigyo Soran* (KSK) publication for 2004. The LME population has a total of 2,748 overseas affiliates in operation for which dates of establishment are given for an average of 6.4 per LME as against Toyota's 91. The LMEs also have 330 representative offices which apparently Toyota foregoes. The affiliates are then defined further as either production or sales operations.

Table 12.4   **LMEs vs. Toyota: Overseas Establishments**

		Overseas Affiliates	Production	Sales	Rep Offices
429 LMEs	Total	2,748	1,437	1,311	330
	Average	6.4	3.3	3.1	0.7
Toyota Motor		91	40	51	–

Source: KSK (*Kaigai Shinshutsu Kigyo Soran*), 2004.

A production affiliate is involved with any aspect of production including manufacturing, processing and assembly, although in many cases this also embraces a sales arm. Production can entail the complete finished item or the manufacture of materials and components. A sales affiliate, on the other hand, is more precisely defined as a non-production establishment because, although in most cases this signifies a designated sales operation, it also covers R&D, information gathering and assessment, and various managerial and financial functions.

The averages tell the difference as to what globalization means for the LME population as opposed to Toyota. The latter is a globalizing company in the full sense of the word because it more or less covers the globe. It can go anywhere it pleases. Whereas, as we shall see below, the LME population as a whole is represented in most of the global territories and to the same degree as Toyota, however, the individual LME is much more circumscribed. It inevitably has to be more selective as far as overseas locations are concerned; it is a firm operating within an environment with global potential if tackled with strategic acumen rather than one which is itself bestriding the planet. Although here again, as demonstrated by Table 12.5, within the LME population itself the hierarchical continuum is once more in evidence. An LME listed in the First Section is likely to have over twice as many overseas establishments as an unlisted LME.

Geography and time

The point that is persistently apparent in reviewing the following data about the foreign direct investment initiated by Japanese LMEs over the past half century is that it is true to a Japanese pattern. That is to say that, despite their singular character as specialist leading medium-

Table 12.5   LME Overseas Establishments along the Continuum

| | | Overseas Affiliates | | | Rep |
		Production	Sales		Offices
248 1st	Total	2,026	1,022	1,004	228
Section LMEs	Average	8.2	4.1	4.0	0.9
121 2nd	Total	517	307	210	67
Section LMEs	Average	4.2	2.5	1.7	0.6
60 Unlisted	Total	205	108	97	35
LMEs	Average	3.4	1.8	1.6	0.6

Source: As for Figure 13.3.

sized enterprises, there is nothing that sets them apart from other Japanese manufacturing firms in terms of timing of overseas investment and geographical spread. To take the latter first, their investments are preponderantly in East Asia, North America, and Europe. East Asia can be broken down into what have been dubbed since their take-off the Asian New Industrial Economies – Hong Kong, Singapore, Taiwan, and South Korea; the original ASEAN 4 of Thailand, Malaysia, Indonesia, and the Philippines, plus latterly Vietnam; and of course China. North America is dominated by the United States and Europe chiefly comprises Britain, the Netherlands, Germany and France, recently joined by the Czech Republic among others. As for the rest of the world, Latin America means Brazil and Mexico, Australia features from time to time, India is now on the visible horizon, and South Africa may be getting there. Put another way, about 15 countries have been the recipients of the vast bulk of FDI by Japanese manufacturing LMEs as they have internationalized.

Looking at Figure 12.1, the geographical distribution pattern for LMEs is similar to that of Toyota. However, the emphasis on East Asia including China as compared with North America and Europe by the LMEs is somewhat stronger than Toyota's, explicable in China's case especially by the fact that there are only so many units a single firm like Toyota needs to establish in a single economic area. On the other hand, Toyota is more extended. 'Europe' includes a fast-growing number of East European bases, while 'Other' encompasses manufacturing in India and South Africa, as well as a bevy of affiliates in Peru, Venezuela and other South American countries where apparently none of the LME sample has invested.

As for timing, Taiwan, Hong Kong and Singapore were the first off the block from the mid-1960s, both for cheaper manufacturing and chiefly local sales, to be followed by South Korea in the 1970s. Simultaneously, the huge U.S. market proved irresistible to many, with sales affiliates springing up at over twice the pace of production units. By the 1980s Europe was picking up steam as a market and it was the turn of the ASEAN 4 for manufacturing. This surge of production investment in Thailand, Malaysia, Indonesia and the Philippines progressed into the 1990s (taking on Vietnam in the process), at which juncture it was first complemented by, but then rapidly and unceremoniously superseded by, the China tsunami hurtling irresistibly through that decade and into the new millennium.

As implied in Figure 12.2, moreover, each of the major regions of North America, Europe and East Asia took on their own distinctive characteristics. While the United States has always generated more

Figure 12.1 LMEs vs. Toyota: Global FDI Spread

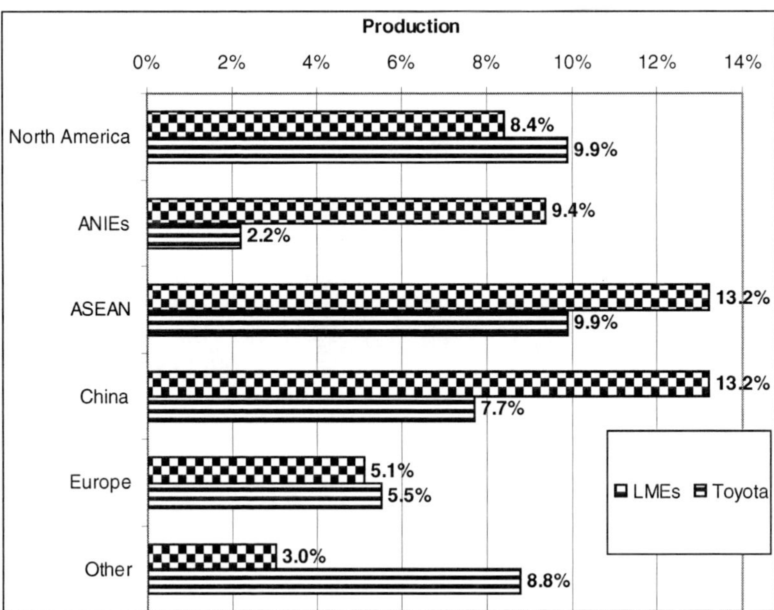

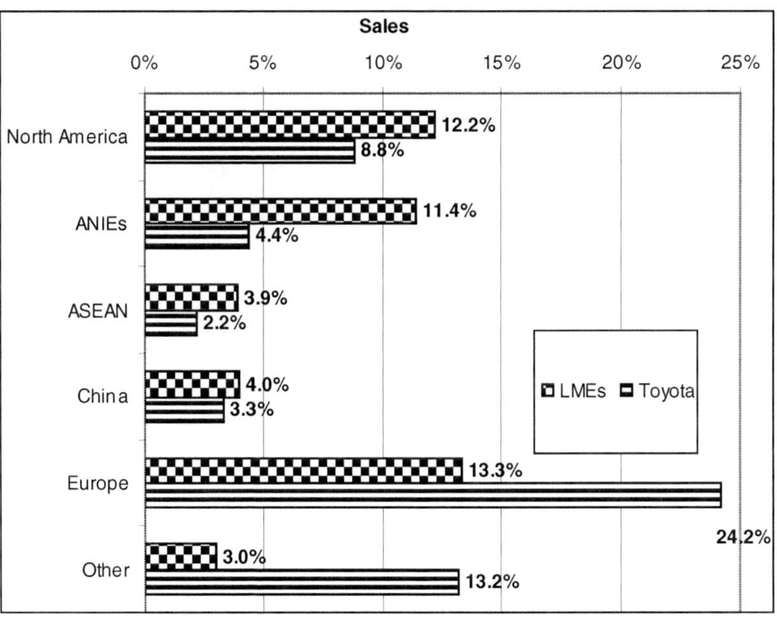

Figure 12.2   Timing and Geographical Location of Overseas Investments

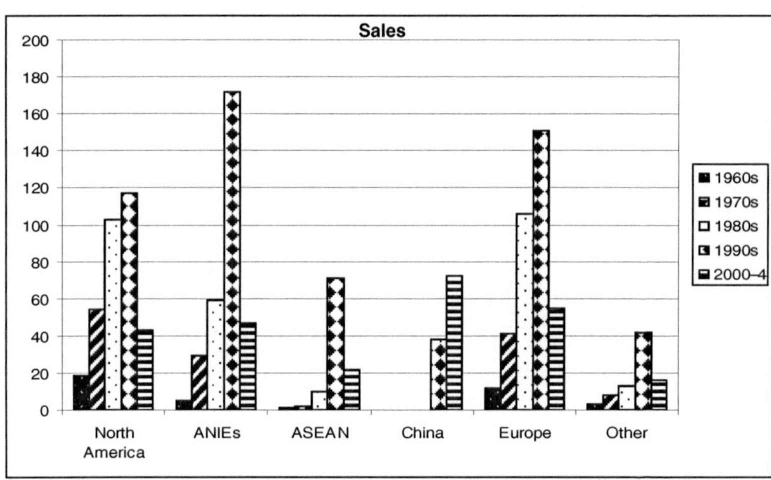

sales affiliates, the number of production units and the consistency over time of their establishment are nevertheless impressive, reflecting the need of these Japanese LMEs to ensure prompt response to demand and reaction to developments in the world's pre-eminent economy. Proportionately speaking Europe has been more weighted to sales, although new production affiliates have steadily increased, starting with Britain and the Netherlands, then Germany, France and Belgium, followed from the 1980s by Spain and Italy, and now increasingly by the Czech Republic, Hungary and Poland. By far the most variegated of the three regions has been East Asia. Having commenced as manufacturing bases, already by the late 1970s Hong Kong and Singapore were coming to be seen as sales entrepôts and logistical hubs. And while their fellow Asian NIEs, Taiwan and South Korea, lasted longer in the production camp due to fast upgrading of technical capabilities and their comparatively more substantial economies, come the 1990s they too were becoming host to proportionately more sales affiliates aimed at their maturing markets. The same pattern is beginning to be discernible for the ASEAN countries, notably Thailand, while even China, after being insistently production-oriented for two decades, seems to be just starting to follow suit. Moreover, it should be noted that as things stood in 2004, China had simply caught up with the other East Asian countries in terms of total functioning investments. It was, in the main, complementing more than replacing.

The East Asian specialist manufacturer

In fact, the combination of complementarities and proximity is an important aspect of the dealings of Japanese LMEs with East Asia. Looking back to the early 1960s, an internationally-minded LME like Shimano as highlighted at the end of the last chapter could see the United States as the 'must' market. Then there were small East Asian markets which could – with the exception of Hong Kong and Singapore – only be cracked by working from the inside because of excessively high import barriers. Brazil featured for some, encouraged by the large Japanese immigrant population, and maybe Europe. Things began to alter through a combination of a changed perception on the part of East Asian governments, starting with Taiwan, in favor of export-led rather than import-substitution growth, subsequent competition from these sources, and mounting trade pressure emanating particularly from the United States. Manufacturing had to be dispersed either to where the markets were in North America and Europe, or 'locally' to the nearby East Asian countries. By 2004 it is quite apparent that the latter absorbed most of this load.

This is illustrated in Figure 12.3. Of the sample 429 LMEs, 276 of them – almost two-thirds – belong to just four of the industrial categories: chemicals, general machinery, electrical machinery, and transport equipment. This group of companies in 2004 had a cumulative total of some 1,012 operating production affiliates worldwide for an average of 3.7 per LME. Almost 70 percent of these affiliates were located in East Asia, that is the four Asian NIEs, the ASEAN 4 plus Vietnam, and China. This is complementarities and proximity at work. The Japanese LMEs, with their

Figure 12.3   Production Locations for Japanese LMEs in Chemicals, General Machinery, Electrical Machinery and Transport Equipment (1960s–*circa* 2004)

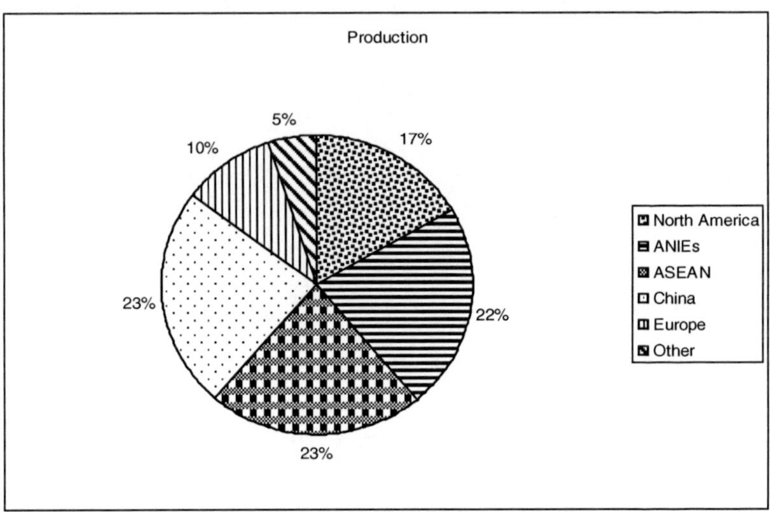

advanced, specialized technologies have needed cheaper production bases, preferably close at hand, to satisfy market demand; the East Asian economies have needed Japanese LMEs possessed of these advanced, specialized technologies to enhance their industrial development. The location percentages remain consistent across the continuum of LMEs except for one factor. That is to say, the United States remains in the 15 to 20 percent range and East Asia stays just shy of 70 percent. However, according to the data, LMEs listed in the First Section of the stock exchange are twice as likely to have production facilities in Europe than the Second Section and unlisted LMEs, suggesting by default as it were that proximity still influences the choice of FDI location, especially for the smaller firm.

This is not unique to Japan, needless to say. When Lego started on its internationalizing trail it established sales affiliates in Germany, Switzerland, France, the U.K., Belgium and Sweden in the 1950s before heading out to North America, Australia and Asia in the 1960s. Germany again was where its first foreign manufacturing plant was inaugurated in 1964. This was a regionally incremental approach in keeping with the times. Now it produces in China, as does Italian brake disc maker Brembo. And times have changed to the point that, rather than being regionally incremental, satellite developer Gilat's approach in the 1990s was globally inclusive. That having been said though, being globally inclusive as a tendency has always been stronger among Western-oriented specialist LMEs than their Japanese homologues. Having come out with the world's first electric powered chainsaw, over the following 80 years Stihl has systematically taken it around the world. To be sure, Pentel and Shimano proceeded in short order to the other side of the world, Nidec went straight to the world's premier clients, and Shofu joined the select fraternity of global initiators with the world's first dental spherical amalgam as long ago as 1965. However, globalization for most Japanese LMEs has tended to be more channeled to specific markets than their Western counterparts. Such channeling, moreover, has been much more markedly apparent in their selection of overseas manufacturing sites, particularly regarding its concentration in East Asia.

The reasons for this are complementary. First, for one and a half centuries Japan built up a demonstrable industrial lead over its East Asian neighbors. In doing so, among other things, it spawned a host of specialist manufacturers, many of which by now are seasoned practitioners. Second, faster than other parts of the underdeveloped world the countries of East Asia have accommodated themselves to industrialization. However, their initial tardiness has created a technological gap. Although this gap can be foreshortened with respect to the assembly of even ultra-modern products, it is far more difficult to close with respect to the inner workings, the key components and materials. This is where the Japanese specialist LME often comes in. It is the reason why giant South Korean assemblers, let alone producers in Malaysia and Thailand, still buy machinery and equipment from Japanese specialist manufacturers. It is why many of these Japanese specialists are now lodged into the East Asian industrial structure, servicing what is increasingly becoming a 'domestic' East Asian market. To proceed a step further, moreover, it is why a company like machine tool maker Makino Asia, based in Singapore and headed by a Singaporean CEO, can expand 'almost five times over a span of six years' in selling its milling machines worldwide as an East Asian manufacturing affiliate of a Japanese LME (EDB, 2008).

Complementarities and proximity: Japanese LMEs and Taiwan

Another way to observe how this combination of complementarities and proximity has functioned is to take the progress of a particular East Asian host economy and match it against the time of entry through FDI of a given selection of LMEs. Taiwan is the ideal example for a number of reasons. First, it is one of the two Asian NIEs of reasonable economic size, being currently 19th in the world in terms of gross national product. Second, it was a good decade ahead of the other large Asian NIE economy, South Korea, in adopting an export-led growth policy incorporating a fairly liberal, albeit carefully monitored and selective, accommodation with inward FDI. Third, the proximity factor is very much to the fore both geographically and culturally. Taiwan having been for 50 years to 1945 a colony of Japan under whose relatively benign tutelage the island developed tolerably well, Taiwanese businessmen overall have taken a constructive and non-antagonistic view as to what Japanese companies can contribute to their own success. Fourth, since the 1960s Taiwan has continued to progress at a fast clip economically and technologically. This has done much to ensure the continued commitment of Japanese investors, despite the emergence of a host of new rivals, including China, and the political uncertainties attendant upon China's claims on the island.

And so to what the Japanese LMEs did and why. When Hisamitsu Pharmaceutical decided at the beginning of the 1960s that Taiwan constituted a market worth exploiting it had to set up shop there. It had no other option because of prohibitive duties in an economy driven by the principles of protection and import substitution. By the time DC motor maker Mabuchi arrived in 1969, however, the atmosphere was very different; it was outward-looking and intent on export expansion. Mabuchi was welcome as a supplier of small electronic motors to power the toys and other simple devices a multitude of Taiwanese micro-firms were starting to churn out. Another appealing aspect for Mabuchi was the cheap and abundant labor which helped keep it competitive, and this factor was likewise on the mind of Janome when it brought its low-end sewing machine manufacturing there in the same year. However, even at this early stage it was becoming apparent that the labor pool was not just cheap and abundant but relatively well-educated to boot, while the proactive administration was making no bones about upgrading and elaborating the industrial structure through the selective insertion of manufacturing capacity and knowhow that Taiwan was as yet in no position to provide for itself. So the end of the 1960s also saw the inauguration of relatively sophisticated inputs by way of, for example, Nippon Piston Ring's engine parts, OSG Corp's precision cutting tools, and Mitsumi's polyvariable condensers.

By the 1970s, a combination of higher land prices and wage costs and stricter regulations at home plus mounting trade friction abroad, as symbolized by Richard Nixon's imposition of much stiffer duties on a whole range of Japanese goods early in the decade, was beginning to persuade a growing number of producers that manufacturing beyond

Japan's shores to serve third markets made sense. One such producer was Daiwa Seiko, whose initial establishment as a maker of sports fishing tackle in the late 1940s, was specifically aimed at the club of advanced countries of which Japan was not a member at the time. From the outset of the investment in 1971, therefore, Taiwan was seen as a complementary export base. But, again, in this fast-moving train of events, as the Taiwanese economy pressed inexorably forward different LMEs with different agendas had their own particular take on Taiwan. Come 1980, Taiwan was once more assuming the character of a market, and not just for intermediary goods, with an economy already very different from that perceived by Hisamitsu in 1961. 1980 was the year, therefore, that Fujitec set up to produce and sell elevators in Taiwan, aimed at a domestic market it deemed sufficiently advanced to support such a venture.

What had happened was that the shift in direction to export-led growth in 1964 had hardened by the 1970s into industrial policy which targeted certain industries, notably electronics. This progress was afforded symbolic expression with the establishment of the Hsinchu Science Park. Also set in motion was an aggressive program aimed at encouraging top-class Taiwanese technicians and managers to return home from California where they had made their mark. During this process, Futaba (electronic devices and components), Daishinku (crystal oscillators) and Nippon Chemi-Con (condensers), who had already expressed their appreciation of the potential thus presented by setting up for manufacturing in the 1970s, were joined in the 1980s by the likes of Idec Izumi (control equipment) and Hokuriku Electric (resistors) as the opportunities multiplied. These LMEs were, in other words, by this stage coordinating with an economy which was becoming increasingly self-driven. Such evolution has meant moreover that, whereas at the start of the process in the 1960s and 1970s, such Japanese LMEs were the essential suppliers of the components and machinery, by the 1990s technological progress among Taiwanese indigenous firms was such that the relationships with them could be much more reciprocal. Thus did I-O Data Device arrive in the capital of Taipei in 1996 not only to sell its PC peripheral equipment, but also to *procure* Taiwanese-made parts.

In addition, for some LMEs Taiwan has acquired a strategic significance within a wider arena – Chinese, regional or global. Not a few of them, like PCB drill bit maker Union Tool, having been established for some time in Taiwan have dispatched their skilled and experienced Chinese-speaking Taiwanese staff to Mainland China to assist in the inauguration and running of facilities as they come on line there. As for regional strategy, by the time Uni-Charm, the maker of sanitary napkins and paper diapers, turned up in Taiwan in 1985 the economy was booming. Given this state of affairs, Taiwan had been consciously selected as the company's first investment in manufacturing abroad because of its perceived suitability as a testing ground for a regional strategy which has since incorporated the whole of East Asia. Taking this one big leap further in 1987, from the outset the manufacturer of mineral insulated cables Okazaki Manufacturing looked on its plant in Taiwan as an integral part of a global strategy based on the technical capabilities of the labor force coupled with the possibility of production at somewhat lower cost than in the United States or Japan. By the new millennium it was claimed that this plant was the world's leading producer of thermocouple mineral insulated cables. Thus did the advanced *monozukuri* of the Japanese LME combined with the emerging skills of a fast-developing economy to global effect (Evans, 2003:209–19).

Machine tools

Encountering globalization

Table 12.6 presents the FDI records of eight machine tool LMEs, five of which have already been encountered in Chapter 7. The other three were founded in the postwar period. They are Mori Seiki, Makino Milling Machine and Obara. Looking at the table, companies seem to wake up to the potential of foreign direct investment at certain times and then engage in bursts of activity, as did Koike in the 1980s, OKK in the first half of the 1990s, and Okuma in the second half of the 1990s and into the current century. In doing this they have been, as a group, highly selective in where they have placed their FDI. East and Southeast Asia, the United States and Europe have taken the bulk of these investments, the Asian areas and the United States being favored for manufacturing, while Europe, especially via Germany, has been serviced exclusively by designated sales facilities. The United States was an early choice for manufacturing, although this is also backed up by sales subsidiaries. The more industrially advanced Asian countries – Taiwan, Singapore and South Korea – were all relatively early venues for manufacturing FDI, typically joined by Thailand in the 1980s, and then by China from the mid-1990s and into the 2000s. Company globalization strategies also vary. Mori Seiki, for example, is highly active overseas but has no affiliate listed as a manufacturer.

Table 12.6   FDI Locations of Eight Established Machine Tool LMEs

Year	Company	FDI Recipient Manufacturing	Sales
1971	Takisawa	Taiwan	
1972	Okamoto	US	
1973	Okamoto	Singapore	
1978	Makino		Germany
1980	OKK		US
1981	Makino		US
1982	Koike		Netherlands
	Mori Seiki		Germany
1983	Mori Seiki		US
1985	Koike	US	
	Mori Seiki		US
1987	Koike	S. Korea	
	Makino	Singapore	
	Obara	S. Korea	
1988	Okuma		Germany
	Okamoto	Thailand	
	Mori Seiki		Singapore
1989	Takisawa		UK
1990	OKK	Thailand	US
	Mori Seiki		France
	Mori Seiki		Spain
	Mori Seiki		Italy
	Obara	Malaysia	
1991	OKK		Germany
	Mori Seiki		Taiwan
1992	Okamoto		Germany
1994	Obara	China	
1995	Okuma	US	
	OKK	China	
	Mori Seiki		Thailand
1996	Mori Seiki		Hong Kong
	Mori Seiki		Brazil
	Obara	Thailand	
	Obara	US	
1997	Okuma	Taiwan	Brazil
1998	Mori Seiki		Mexico
2000	Okuma	Thailand	
2001	Okuma		China
	Mori Seiki		China
	Obara	China	
2002	Okuma		Australia
	Koike	China	
	Takisawa	China	
	Mori Seiki		S. Korea
	Makino		Singapore
	Obara	Mexico	

Table 12.6   FDI Locations of Eight Established Machine Tool LMEs – *continued*

Year	Company	FDI Recipient Manufacturing	Sales
2003	Okuma	China	
	Takisawa		US
	Mori Seiki		Indonesia
	Mori Seiki		Germany
	Mori Seiki		Australia
	Obara	Australia	
2004	OKK	China	
	Mori Seiki		France
2005	Okuma		New Zealand

Sources: KSK and the individual company websites.
NB: 1. 'Manufacturing' also includes sales, while 'sales' is exclusively sales.
   2. In addition to manufacturing and sales facilities, companies may have
      representative offices which are not recorded in this table.

Obara, by contrast, turns up with plants wherever the automobile industry is taking root. It is a prime example of one of those smaller companies where a degree of specialization has been attained in a narrow field of expertise. It is a postwar startup commencing operations in 1958 with the manufacture of electrical contacts. Over the years since it has established itself as the leading manufacturer of the resistance welding machinery, most of it destined for the automobile industry, and the company commands 50 to 60 percent of the Japanese market. In addition, Obara now boasts manufacturing bases in the United States, China, South Korea, Malaysia, Thailand, Mexico, and Australia, and has a global market share of close to 20 percent. By 2005, Obara was in full swing as the Japanese and Korean auto makers indulged in a revitalizing investment spree, contributing considerably to the orders both at home and abroad, which would ensure Obara its highest profits ever for the period ending September of that year (Nikkei, 2005c). Resistance welding entails the application of a high amperage electrical current to a combination of two or three steel sheets held in place by several hundred kilograms of force so that they are welded together by the heat generated by the current. This allows for welding at low cost in a short time and has thus become the main form of welding for automobiles (www.obara.co.jp; Nikkei, 2005d). Doing this better than the competition is an essential ingredient in Obara's globalization strength.

Success in the 21ˢᵗ century

Japan's economic recovery got under way in earnest as of 2004 and many LMEs have been recording substantial profits since then. The financial press has been replete with incidences of historical highs, evidence of new and expanding markets, notably Eastern Europe and the BRICs – standing for Brazil, Russia, India and China, as well as of intensified attention to efficiency and productivity. LMEs in the machine tool industry encapsulate the process and provide an instructive illustration not least because they do not so much represent ultra modernity but rather a mature industry in its prime. These Japanese LMEs are not unusually world leaders in their core competence, capable because of their accumulated skills and experience of state-of-the-art solutions which few if any can match. Mature and accumulated are apt epithets here as well because, apart from a few extra entrants in the 1970s mainly associated with the numerical control revolution, the structure of the industry in Japan and the positioning of the companies within it were essentially fixed by the end of the 1950s. This was the phase, as we have seen, when the government was highly solicitous of the machine tool industry's welfare which galvanized a surge of entrepreneurial vigor. But early government exhortation and support aside, what we see in the middle of the first decade of the 21ˢᵗ century is the triumph of the artisan in his implacable application of *monozukuri*.

To begin with, notwithstanding the indisputable contribution of exports to the resurgent vitality of the Japanese machine tool industry right now, domestic demand has also had much to say about what has been happening. By 2004, for instance, many major Japanese manufacturers were redirecting a significant proportion of their manufacturing back to Japan, on top of the fact that they were exporting more anyway. This led to a rush of facility expansion and plant construction which inevitably translated into a bonanza for the machine tool makers, pressing them likewise to expand to meet urgent deadlines and clear the backlog of orders rapidly approaching in magnitude the halcyon days of 1990, just prior to the bursting of the bubble. In fact, by 2006 the Japan Machine Tool Industry Association was able to report that, in amounting to some 720 billion yen in value for the first six months of that year, orders had exceeded those for the equivalent period in 1990 by 2.3 percent (Nikkei, 2006c).

The result is that by 2007 the Japanese machine tool industry taken overall was enjoying an unprecedented golden era. An interview given at the end of February by Hanaki Yoshimaro, president of Okuma, fills out the details. He notes that the value of machine tool production in

Japan for the previous year hit ¥1,437.1 billion, surpassing the previous peak achieved in 1990 for a new historical high. Moreover, by November of 2006, orders had increased for 50 successive months on a year-on-year basis. Such was the demand that the only major problem the industry confronted was the time needed to train technicians to the required high standard, although meeting order deadlines also presented an enormous challenge. Exports as a ratio of this demand have grown dramatically over the one and a half decades or so since 1990; in January of 2007 they were already exceeding domestic deliveries and looked set to continue to do so throughout the year, whereas they only accounted for 25 percent of total shipments in 1990. The emerging economies of China and India are top of the client list, joined by the Czech Republic and Hungary, and more recently Romania, Bulgaria, and Slovenia.

In terms of the many industries being served, demand turned positive for everything from automobiles to electric machinery and electronics to resource development. And while the competition from the likes of South Korea, Taiwan and China is also growing to meet this expanding worldwide demand, the Japanese still command the lead in the high-performance, high value-added arena, a lead which if anything is widening according to Mr. Hanaki, an engineer first but with a wealth of sales experience as well. This being so, Japanese manufacturers need not be overly concerned about the global competition, he concludes (Nikkei, 2007e). They will continue to contribute to globalization in three ways: first as suppliers to domestic exporters, second as exporters in their own right, and third as overseas manufacturers.

Summary

By 2004 the long recession had ended and Japan was reasserting itself. While globalization seemed more manageable, however, there were new challenges in the form of emerging economies to encompass as well as encroaching alien practices to contend with. A population sample of 429 LMEs is used here to follow their course of internationalization up to 2004. To reconfirm what the LME is Toyota is employed by way of contrast. Although it is huge by comparison it still has some similarities with the LME, including a family presence and a devotion to *monozukuri*.

The LMEs in the sample are both listed and unlisted. They are recognized leaders in their elected fields, have domestic workforces of 200 to

2,000, are independent and not subsidiaries, and have track records overseas. In terms of capital stock and number of employees the average LME in the sample is less than two percent the size of Toyota. A distinction is made between family and non-family firms and, in the case of listed companies, those in the first and second sections of the stock exchange. This is to illustrate the continuum described by them, which is also illustrated with respect to the listed firms by the stock percentages held by banks and trusts, notably Master Trust Bank of Japan and Japan Trustee Services Bank.

In the sample there are 265 family LMEs and 164 non-family, or alternatively 369 listed firms and 60 unlisted. The two predominant industrial categories are general machinery and electrical machinery, these being followed by chemicals and transport equipment. By 2004, these LMEs had an average of 6.4 affiliates abroad as compared with 91 for Toyota. These then are further defined as production and sales affiliates. Japanese LME FDI follows the same pattern as that of Japanese firms in general, in that the vast bulk of it is accounted for by about 15 countries in North America, Europe and East Asia. However, that of Toyota is spread wider to take in Eastern Europe and South America for example. In terms of timing, the LME FDI first went to the Asian NIEs for production and the United States for sales and some production. By the 1980s it was in Europe for sales and ASEAN for production, before pouring into China for production in the 1990s. About 70 percent of the production affiliates are now in East Asia, where a combination of complementarities and proximity has been at work. To that extent East Asia is assuming the character of a 'domestic' arena in the strategic mix of many of these firms. This is best illustrated by Taiwan which has been in turn for Japanese LMEs protected consumer market, supplier of cheap labor and buyer of inputs, production and export base, burgeoning consumer market, hi-tech collaborator, and global partner.

13
2007 Update: Funds and China

Portents

The sample analysis in the last chapter runs to 2004. The period of three years since then to the time of writing is hardly long enough to distinguish with absolute certainty the emergence of new trends in the world of the Japanese LME, but enough has occurred nevertheless to enable the observer to be conscious of significant portents, signs implying the course of transformations in the offing. If they have a leitmotif it is the emerging foreign factor; the choices presented to Japanese LMEs are increasingly laced with an international flavor. Of the LME population of 429, five were listed for the first time on the stock exchange during the period from 2004 to 2007 while another 15 were upgraded within it. Over the three years in question no sample LME simply went bankrupt. However, a handful were delisted having been bought out or merged with companies in the same industry and of similar size as has already been recorded in Chapter 9 regarding toy makers Tomy and Takara. Most of this merger and acquisition activity was amicable, but there are omens that the alien hostile takeover is on the horizon. There was also one management buyout – another recent import – involving the soft drinks maker Pokka for reconstruction purposes. This case and Takara Tomy also point to another recently developing scenario: the involvement of funds in corporate management as shareholders. For Pokka and Takara Tomy they are welcome partners. For others they are aggressive interlopers and often, moreover, not Japanese. This growing foreign involvement in ownership together with the irrepressible rise of China are the two factors most likely to affect the destiny of Japanese specialist manufacturers over the coming decade.

Ownership and the foreigner

It is the foreigners who have been making the waves over the past three years. Not only has the stake taken in Japanese companies by foreigners as a whole been following a rising trend, reaching a new high for the fourth consecutive year in fiscal 2006 of 28 percent, but the number of foreign shareholders has been going up nonstop for 11 years, reaching almost 40,000 for the same year (Nikkei, 2007f). In tandem with this, foreign institutional investors have been expanding in number and taking a firmer grip on events. The character of foreign presence and participation is in the process of a substantial transformation. This can be seen even among the staid operators who have been in the field for some time. It has been the custom of the likes of Goldman Sachs, Chase Manhattan and Morgan Stanley to invest in a fair spread of Japanese firms, albeit usually taking a back seat with modest stakes of well under ten percent.

But the veil could be about to drop. Restricting our attention to LMEs and citing from the Autumn 2007 edition of the Japan Company Handbook, we see that Goldman Sachs has raised its stake in the chemical company C. Uyemura to 11.2 percent to become the leading shareholder, while State Street Bank and Trust likewise has assumed the top perch at Shimano with 13 percent. Moreover, more aggressive players are coming to the fore, more overtly intent on making their presence felt in company operations. One such example is Bear, Stearns & Co. Inc., a New York-based investment banking, securities trading and brokerage firm. Bear, Stearns was, as of the fall of 2007, the leading shareholder of precision cutting tool maker OSG with 8.9 percent, Nifco, a major producer of industrial fasteners, with 9.9 percent, and Miraca Holdings, a top manufacturer of diagnostic drugs and equipment, with 9.4 percent. Another major shareholder of this last company is Steel Partners, an investment fund.

The sound of Steel

If it has been the foreigners who have been making the waves over the past three years, it is investment funds which have been making the noise, and responsible for the highest decibels has been Steel Partners Japan Strategic Fund LP. This fund is dubbed a 'shareholder activist', along with, among others, fellow Americans Perry Capital and Dalton Investments, UK-based Silchester International Investors, and the indigenous asset management firms Asuka and Ichigo. As shareholders they are rapidly rising in significance. According to a recent survey,

Table 13.1   Steel's LME Stakes

Company	Product	Exchange	Steel's% Share	Steel's Rank
Ezaki Glico	Confectionary	Tokyo 1st Section	14.3	First
Fukuda Denshi	Medical micro-electronic equipment	Jasdaq	14.0	First
Yushiro Chemical	Metalworking oil	Tokyo 1st Section	13.6	First
Nihon Tokushu Toryo	Paint	Tokyo 1st Section	13.3	First
Nissin Food Products	Noodles	Tokyo 1st Section	9.5	First
Bull-Dog Sauce	Condiments	Tokyo 2nd Section	9.3	First
Tenryu Saw Mfg	Circular saws	Jasdaq	8.1	Second
Shofu	Dental materials	Tokyo 2nd Section	8.1	Second
Miraca Holdings	Diagnostic drugs	Tokyo 1st Section	5.3	Fourth

Source: JCH Autumn 2007 edition, issued on August 25, 2007.
NB: The report in the text stating Steel's stake in Nissin Foods as 18 percent was printed in October, 2007.

12 funds which can be described as activists held shares exceeding five percent in 93 listed companies. Steel had 30 such cases in its portfolio. Table 13.1 gives a list of LMEs in which Steel has stakes of above five percent according to the August 2007 edition of JCH.

The background to this is the evolution of corporate Japan over the past 20 years from one characterized by informal networks of cooperation and guidance to one much more subject to legal interpretation and enforcement. One consequence of this, plus the aggravation of the protracted economic downturn lasting a decade from the early 1990s, was the weakening of ties with accommodating banks as they started selling off their shares in companies. In addition, given the weak economy, firms refrained from investing in it and hoarded cash instead, with a consequence that an increasing number of them became seriously undervalued. Many of these were what have been defined here as LMEs and therefore by definition sound operationally. Their problem, in the eyes of investment funds particularly, is that they are financially unsophisticated and therefore present a substantial market to be tapped. Another consequence has been the more explicit legally stipulated responsibilities and powers of companies, boards and shareholders. Activist funds have adroitly positioned themselves as champions of shareholder rights in this process. As prominent shareholders themselves, they criticize the firms to which they have committed their capital for inefficient use of resources, bloated staff, and the maintaining of unnecessarily large cash reserves, while also demanding higher dividends. This is welcomed by many individual Japanese shareholders who can justifiably claim to have had a raw deal dividend-wise over many years.

However, there is another side. In a sense this is a war of attrition against traditional business practices. Attrition because the companies usually targeted at this early essentially speculative stage are not the largest corporations but the more easily accessible medium-sized firms. This as noted includes LMEs; MBO fund Vestar touched base in Japan in 2006 to specialize in them (Nikkei, 2006d). And attrition also because it is part of a process whereby, not only the different company product departments can be parceled out as discrete entities subsequent to take-over for example, but also the functional units within the firm – board of directors, management, employees, shareholders – are being separated off from each other. The family risks being broken up. What is more, with mounting pressure for the appointment of independent directors, among other factors, the norms of the outsider as discussed with respect to American corporations in Chapter 3 are starting to infiltrate. It would seem that the chamber of our stylized LME is being dented, if not as yet dismantled, and funds are the motors accelerating this trend.

A Japanese response

But the funds are not finding the going easy. New ways of doing things, based as they are on a radical shift in perception as to what a company is, what it stands for, and whom it is for, are finding it difficult to reach *terra firma*. The world may have been undergoing an M&A boom with activity up 50 percent from the previous year, and the number of transactions involving Japanese companies may have shot up over the last decade to reach 15 trillion yen ($129 billion) last year, but the value of the deals in fiscal 2006 was only some three percent of that nation's gross domestic product compared with a typical ten percent in America and Europe (*The Economist*, 2007b:68). That is to say, not only is the art of M&A far from deeply ingrained in Japanese business practice, but also that for many Japanese executives it is just as likely to conjure up images of foreboding than opportunity.

Nevertheless, whether the solutions of these executives are appropriate or not, they are right in seeing the takeover as the issue. The fate of the takeover as an acceptable business practice in Japan – and particularly the hostile takeover – is where the battle line is now drawn. Events have unfolded with great rapidity in fact. Barely having arrived on the scene, the wagging finger of investment fund logic began to decree that if the recalcitrant company would not listen to well-intentioned advice of the fund as its conscientious shareholder the only recourse open was a buyout. And if the company would still not listen to reason the takeover would have to be hostile. Then, if the standard buyout formula was adhered to, the fund would transform its purchase into a much more attractive business proposition, before selling it off wholly or in bits at a handsome profit. Corporate Japan, in the main, does not like this. In a survey conducted at the end of July 2007, 70 of the 100 company presidents asked about their attitude towards hostile takeovers responded 'negative' (*hiteiteki*) (Nikkei, 2007g). To them the firm is not a commodity to be manhandled by the shareholders at the expense of the interests of the other stakeholders – management, employees, suppliers, and customers. Moreover, even though the investment funds have insisted, of necessity given these reservations, that they are not intent on cherry-picking, suspicions run high. For LMEs, then, this could be a testing time. The infiltration of investment funds goes to the heart of what they had assumed to be their *raison d'être.*

The objection is that the cardinal objective of the investment fund is simply to make money, in so doing championing the cause of just one stakeholder, the shareholder. This threatens to stifle the *monozukuri* ethos which has brought many Japanese LMEs to where they are today. *Monozukuri* needs a 'chamber' where it can play out its fantasies and devise its own solutions – as we saw with another self-made leader Union Tool in Chapter 9 – removed from the prying eyes of the technologically uninitiated. Nitpicking profit-oriented overseeing is prone to be unappreciative of this, with the consequence that promising new departures can be nipped in the bud and unfulfilled potential abandoned halfway. In this book the LME has been depicted as an innovative chamber, keen to be successful in every way including financially. LMEs have some very rich founders and inheritors. But the trajectory of the LME predicated on its core competences is much broader than that of the fund. The latter was not there at the creation; it is there now because the creation has succeeded. It has a narrower agenda emanating from a culture with different priorities. It may, given its own references, justifiably claim that Japanese business practices are fuzzy, inefficient, and lacking in transparency. It could well be instrumental in realizing improved results. But if, as seems likely, funds are in Japan to stay, roles have to be strictly defined and surprise maneuvers kept within bounds. That calls for vigilance on the part of the LME and restraint on the part of the fund.

Box 13.1   Bull-Dog vs. Steel

Bull-Dog Sauce started as a food wholesaler in 1902 and commenced making Worcester sauce in 1905. It owes its name, which it adopted on being incorporated in 1926, to that canine symbol of Britain, the country where the sauce originated. It is now one of Japan's leading manufacturers of sauces and other condiments and reported to be financially sound. Steel Partners LLC was founded in 1990 and characterizes itself as a long-term investor intent on cooperating with the managers of the companies in its portfolio for the benefit of all stakeholders. As of the middle of 2007 it had invested about $3 billion in corporate Japan, targeting LMEs in the main which it deems to be trading far below their book value per share. Having failed in hostile takeover bids of Yushiro Chemical in 2004 and Sapporo Holdings earlier in 2007, it turned its sights on Bull-Dog.

The Chronology

2002 – Steel Partners becomes Bull-Dog's top shareholder.

2007

May 16 – Steel Partners announces an unsolicited tender offer for Bull-Dog aimed at acquiring 100 per cent of the voting rights of the company.

May 18 – Steel Partners begins the tender offer for Bull-Dog.

June 7 – Bull-Dog decides to counter the tender offer by introducing a takeover defense system amounting to a poison pill scheme.

June 13 – Talks between the two fail to resolve their differences. Steel seeks a court order to bar Bull-Dog's takeover defense.

June 15 – Steel raises its offering price from ¥1,584 to ¥1,700 and extends the period of the tender offer from June 28 to August 10.

June 24 – At Bull-Dog's AGM the poison pill scheme is endorsed by 88.7 per cent of the voting rights at the meeting, or almost all shareholders except Steel Partners.

June 28 – The Tokyo District Court rejects Steel Partners' request to bar Bull-Dog's takeover defense. Steel appeals the ruling.

Box 13.1   Bull-Dog vs. Steel – *continued*

July 9 – The Tokyo High Court turns down the appeal.

July 10 – Steel appeals again to the Supreme Court.

July 11 – Bull-Dog launches its defense system.

August 7 – The Supreme Court rejects Steel's appeal.

August 9 – Bull-Dog issues the new shares.

August 10 – The period of the tender offer terminates.

November 16 – Bull-Dog announces a deficit in its mid-term consolidated settlement of accounts of ¥1.9 billion. Among others, losses incurred included ¥2.1 billion for paying off Steel and ¥600 million for attendant fees to legal and securities advisers (Nikkei, 2007h).

The poison pill

The scheme entailed Bull-Dog issuing three equity warrants for each of its shares, including those held by Steel Partners. This was in order to dilute Steel Partners' stake in the Japanese company to 2.86 percent from the level at the time of 10.52 percent if the fund did not withdraw its tender offer by a given deadline. The equity warrants, or stock acquisition rights, were to be convertible to new shares by anyone but Steel Partners. Instead, the warrants Steel held would convert to a cash payment of 396 yen each. In effecting the scheme, Bull-Dog became the first company in Japan to adopt a takeover defense aimed at diluting an aspiring buyer's stake. Moreover, unlike the conventional takeover defense system aimed hypothetically and in advance at any unwanted investor imposing, Bull-Dog's was only applicable to this single tender offer, specifying Steel Partners as the hostile bidder. It was able to come up with its particular version of the poison pill due to the massive AGM support.

The counterclaims

The objections voiced by Steel Partners with respect to Bull-Dog's defense system are essentially fourfold. First, the defense system discriminated against a single shareholder, Steel Partners itself, and so breached the principle of shareholder equality, while being harmful to both corporate value and shareholder interests in general. Second, the action by Bull-Dog reflected a misunderstanding of Steel Partners' potential role. It was not there to interfere with management, only to participate in supervising with the other directors on the shareholders' behalf to ensure the latter obtained the largest feasible dividends. Its portfolio companies have come to appreciate this, it asserts. Third,

Box 13.1   Bull-Dog vs. Steel – *continued*

being discriminatory, such a poison pill scheme is in violation of Japanese law and so detrimental to the legal framework of corporate Japan. Fourth, the type of defense system Bull-Dog was proposing threatened to undermine international faith in the integrity of the Japanese capital markets and could deter foreign investment in Japanese firms.

The judgment

The Tokyo District Court denied Steel Partners' application on the basis that it did not believe the equity warrants proposed by Bull-Dog were unlawful, nor did the poison pill run counter to the principle of treating all shareholders equally. The court said the issuance of equity warrants should be allowed for three reasons: (i) Bull-Dog passed a special resolution to that effect at its AGM; (ii) all of its shareholders would be appropriately awarded in proportion to the number of shares they owned; (iii) their economic benefits would be secured equally. As long as these three conditions were satisfied shareholder equality would not be violated. Hence, the court relied heavily on the overwhelming AGM approval in judging the poison pill as not discriminatory.

The Tokyo High Court agreed that the planned issuance did not run counter to the principle of shareholder equality and it was fair given the cash payment Steel would be able to receive for the warrants it presented. It was essential and rational, the court opined, for a company to prevent its corporate image from being damaged, even if that meant different treatment of shareholders. In carrying out its business, it added, a company needs to take into account not just the shareholders but all kinds of other stakeholders, including employees, consumers and business partners. Moreover, this court described Steel Partners as 'an abusive bidder' because it ultimately looked to selling off the acquired company assets and was to be seen as solely interested in its own profit.

The Supreme Court backed the lower court rulings. However, although it did not explicitly move to erase the abusive bidder judgment from the lower court record as Steel Partners had demanded, it nevertheless refrained from issuing an opinion of its own. This can be interpreted as the highest court in the land not overtly encouraging the proliferation of poison pill defenses.

Box 13.1   Bull-Dog vs. Steel – *continued*

The president

The above court rulings give something of the flavor of what the company is expected to be in Japan, how it is expected to act, and how it is regarded. A few comments from Bull-Dog's president two months after the curtain came down on the above drama provide further insights.

The AGM approval: 'When we proposed the defensive mechanism at the 2007 AGM in June, I did not imagine we would get so much support.'

The employees' agitation over the events: 'It can't be said that there was absolutely none at all, but people were less perturbed than I thought they would be. Not one person resigned.'

Personal feelings towards the company (having repelled the buyout assault): 'Basically they haven't changed. Shareholders are of course important, but the company comes into being and exists with the approval and support of society. It is essential for it to gain trust by the work it does every day.'

The company and capitalism: 'In working in a capitalist world, the market cannot be denied. If a manager better than me came along there would be nothing for it but for me to step down. The mission of a company is to get results and enhance its contribution to society.' (Nikkei, 2007i).

Interesting in itself one might say. But this president embodies one more portent which will almost certainly take longer to be realized on any meaningful scale than even the hostile takeover. The president, with whom Steel contended and who has worked for some 40 years through the company's ranks to gain that position, is a woman – Ikeda Shoko, a rare phenomenon indeed in present-day corporate Japan.

China

A new monozukuri

For 15 years to the middle of this decade, the overriding international concern of corporate Japan was the rise of China and the threat it could pose. To be sure, through most of the 1990s with respect to quite a few executives and administrators this was more of a platitudinous projection of some vaguely adumbrated eventuality. After all, China was a planned economy controlled by a Communist government, and its giant state-run enterprises were wallowing in debt. There was little chance of China challenging Japan's technological superiority over a broad front for quite some time. Moreover, Chinese workers may be cheap, but they lacked a collaborative spirit and simply performed the tasks allotted to them. In short, Chinese industry was populated by cumbersome dinosaurs staffed by automatons.

But then shockwaves began to ricochet across this easy complacency. It was suddenly apparent that China was a leading exporter of televisions, microwave ovens, washing machines, refrigerators, motorbikes, and mobile phones. Of course, much of this could be attributed to the foreign affiliates which had set up in China, including Japan's Matsushita Electric, Sony and Honda. But a new perception was also dawning. It was not the 'old China' of state-operated enterprises that was making the running, but the 'new China' of youthful and business-savvy managers and entrepreneurs; those who spearheaded the likes of TCL, Haier and Lenovo. Added to which, the window of opportunity had opened beckoningly to a new generation of Chinese workers, young, poor, hungry for success, and not deterred by failure; the same sort of raw energy, in other words, that had powered Japan's high-growth era. The message was clear: a new wind of *monozukuri* was blowing across the globe from China, a wind which was bound to outdo by far the one that had blown from Japan in the second half of the 20th century (Nikkei Business, 2002).

China's achievement

As this growth progresses, moreover, the character of the economy is transforming; the leading force behind the country's export surplus has shifted from textiles to electrical and electronic products. What is more, China is lining up to become one of the top locations for research alongside the United States, Europe and Japan. Beijing has its own Silicon Valley, Zhongguancun, which encompasses among others the research institutes of the universities of Beijing and Qinghua and the Chinese Academy of Science as home to an academic superelite.

This is a far cry from when farmer's son Lu Guanqiu was hired by a local village chief to run a tractor-repair shop in the late 1960s, having started on his own with bicycle repairs earlier in the decade. Mr. Lu is now the multi-millionaire chairman of Wanxiang Group Co. with some 30 ventures worldwide, 40,000 employees, and an annual turnover of over US$4 billion (Wang, 2005). He is the supreme example of one of the major factors which have driven the nation's economy to where it is now: industrialization through the back door launched by farmers in the teeth of the opposition of most of local and central communist officialdom – in the first instance at least. At any event, by 1979, the year after China started opening up and experimenting with capitalism, Mr. Lu turned the repair shop into an auto-parts factory *en route* to producing universal joints and ultimately most of the components in the underside of a car.

A generation on and another factor has emerged as personified by Wang Chuan Fu. If Mr. Lu comes from the school of rolled-up shirt sleeves and callused hands, 'the Chinese farm kid who has gone from the fields into the world' as the company website has it (www.wanxiang.com.cn), Mr. Wang is the representative of talented young individuals capable of dealing with the modern world because they have received their education and training as the market economy has evolved and have built up experience in that environment from the outset. This is why in 1995, while still in his late 20s he could found BYD, which is now one of the world's leading manufacturers of lithium ion and other secondary batteries. BYD's strength lies in the high level of technological capabilities coupled with a cost competitiveness said to make it 40 percent cheaper than its Japanese rivals. The telling point here also is that, according to Mr. Wang, when they started they looked to buying technology and equipment from Japan, but found them unreasonably expensive and so developed them themselves (Nikkei Business, 2002:33–4).

Other factors pertinent to China's meteoric rise are these. First, China made itself available to the global economy at a timely stage in the furious advance of foreign direct investment. As we have seen in the last chapter, China presented the next port of call after the Asian NIEs and the ASEAN 4 and, initially, a lower bottom rung to squeeze costs even further as the international division of labor was elaborated in response to intensifying global rivalries. Second, progressing in tandem with China's incremental engagement with the outside world have been information technology and modularization, both of which, in being cardinal elements in the elaboration of the international division of labor, have proved complementary for a country with a huge pool of largely unskilled workers. Modularization lowers the barriers for entering manufacturing processes and allows for geographical dispersion to the optimal location for the production of each component module and the parts of which it is comprised. It has progressed with personal computers and household appliances and is spreading to automobiles and other industries where China has achieved, or is likely to achieve in short order, a prominent world position. A third factor is the multitude of mainly unprofitable firms producing the same items in large volume and engaging in cutthroat price wars intent on stifling the opposition come what may. Although this kind of activity contravenes the niceties of market principles, it has nevertheless been *de rigueur* for late developers particularly in East Asia. When indulged in on such a scale as China's it can have a formidable impact on trade. It is responsible for Chinese motorbikes finding their way to Venezuela and Columbia, for example. While many may go under, the survivors are establishing beachheads often in markets ignored by advanced country firms, from whence they will be difficult to dislodge.

The road to the 'smiling curve'

How Japanese LMEs fit into this can be viewed from the angle of what China needs. Again we are back to the Japanese specialist as contributor to East Asian development. China may have over the past 25 years raised its share of world exports from 1.8 percent to over seven percent to now occupy second spot, but it has also become the world's third largest importer and one of the top FDI recipients. This is because China is in effect a specialist in finished goods such as apparel, computers and hi-fi video equipment. Although things are changing and, for instance, Japanese household appliance manufacturers based there claim now to be buying more parts produced by indigenous firms than they do in Southeast Asia, the Chinese remain nevertheless uncompetitive over a wide range of more sophisticated parts and components, as well as machinery and equipment. Consequently they have to import them or induce foreign producers to set up locally. China has become the fulcrum of a supply chain, whereby 40 percent of its imports, 60 percent of which are from the Asian NIEs and Japan, are capital goods and intermediary goods to be assembled into products mainly for export. From Japan, by way of illustration, China imports semiconductors, electronic parts, machine tools, auto parts, sheet steel and a wide range of other industrial goods and components. Moreover, many of these sophisticated inputs are destined for foreign affiliates, underlining the phenomenon most notable in the electrical and electronics industries of made *by Japan* or *by Taiwan* in China.

While expediting China's growth, then, modularization and standardization could stunt its innovativeness and increase its dependence. Foreigners account for 66 percent of the patent applications in China and 30 percent of patents applied for abroad for inventions realized in China have foreign co-inventors, as opposed to, say, 3 percent of the cases of patent applications made by the Japanese in Europe (Missions Economiques, 2006). Much of this can be blamed on the country's development model and the state of education partially stemming from it. The development model is still predominantly based on unskilled labor. This has led to the paradox that although, on the one hand, firms are crying out for adequately technically trained individuals there are few available and they are difficult to retain, while on the other hand graduate starting salaries are extremely low averaging no more than US$250 a month. In other words, the foundations can hardly be said to be in place for the creation of an adequately numerous middle class of engineers, researchers and managers necessary to stimulate a steady flow of incremental innovation and development with an enhanced native input (Rocca, 2007).

Moreover, for people in a hurry like Wanxiang's Lu Guanqiu, upgrading *monozukuri* at an acceptable pace is all but impossible to achieve in-house because the chamber of knowledge and expertise is wanting. Universal joints maybe, but the knowledge of electronics and the high level of engineering skills needed to produce the airbag has not arrived on the home front in sufficient numbers to be worth talking about. The solution for Mr. Lu is to purchase advanced-country production units and R&D facilities (Wang, 2005). Ample hands-on experience though he may have had, this makes Mr. Lu an entrepreneurial manager rather than the product-oriented perfectionist that many Japanese LME founders have been. It also places him as much in a global rather than purely national setting. Chinese *monozukuri*, this implies, is unlikely to be decked out in national colors to the extent that Japanese *monozukuri* so often has been. And what this points to is not just the spaceships that the denizens of Beijing's Silicon Valley may be dreaming up, but also includes the valves capable of resisting ultra-low temperatures for use in rockets made by an LME called Fujikin and Pentel's Sign Pen taken aboard the Gemini flights in the 1960s because it did not leak under extreme atmospheric temperature changes.

This is best illustrated by what is called the 'smiling curve'. This curve shows the value added to a product at each stage, starting high with the conception, sloping down through component production, then bottoming out with assembly before rising once more through sales and after-sales services. Chinese industry is mainly situated at the assembly and therefore lowest value-added stage. This is where Taiwan and South Korea started out too in their early industrialization phases, except they were faced with a gentler curve. With modularization the slopes of the curve have become much steeper and China has been competing with other developing countries to secure the stage of manufacturing where value added is in fact declining. By contrast, the Japanese specialist manufacturer is now repeating a role it has already played with Taiwan, South Korea and the rest of East Asia, supplying the missing links in the supply chain, but on a much vaster scale. As a rough rule of thumb, the LME population used in this book can be characterized as engaged 70 percent in intermediary goods and 30 percent in end-user products. On the smiling curve, which is now steeper than ever, that puts many of these LMEs at the lucrative component production stage. That is one very good reason why they have been pouring into China at an accelerating pace.

Japanese LMEs in China

As we saw in Chapter 12, investment in China by Japanese LMEs was still a mere trickle in the 1980s before taking off in earnest in the 1990s. Figure 13.1 shows what this amounted to in terms of cumulative totals for affiliates established during the period and still functioning at the beginning of 2004. With respect to production affiliates, Greater Shanghai just noses out the Northeast which includes the major cities of Beijing, Tianjin, Dalian and Qingdao. However, immediately north of and contiguous to Shanghai is Jiangsu Province, and these two, combined with Zhejiang Province (one of the larger contributors to 'Other') just to the south of them, account for virtually half of the total Japanese LME investments made during the period as represented by the sample population of 429. Guangdong, although consistently hosting new investments throughout, is already proving less of a pull than Shanghai and its extended hinterland. Furthermore, we are left in no doubt of the overwhelming attraction of that city when we look at the sales affiliates – bearing in mind, of course, that this includes other functions such as importing, marketing, R&D, and financing. Over three quarters of the LME population non-production affiliates in China are based there. The location choices, timing and purposes of the production affiliates of just one company can illustrate the evolving dynamics at play, as they pertain to both China and Japan.

Mabuchi has been for some time now, to quote the current chairman Mabuchi Takaichi, 'the world's leading manufacturer of small electric motors'(Mabuchi Motor, 1999). Being heavily reliant on inexpensive labor it has also played the China card to the hilt, starting very early. It was already in Hong Kong in 1964, only ten years after its establishment, thereby being a pioneer in its industry in the implementation of international division of labor. The motors it produced were for the

Figure 13.1   LME Affiliates Established in China, 1980s–2004

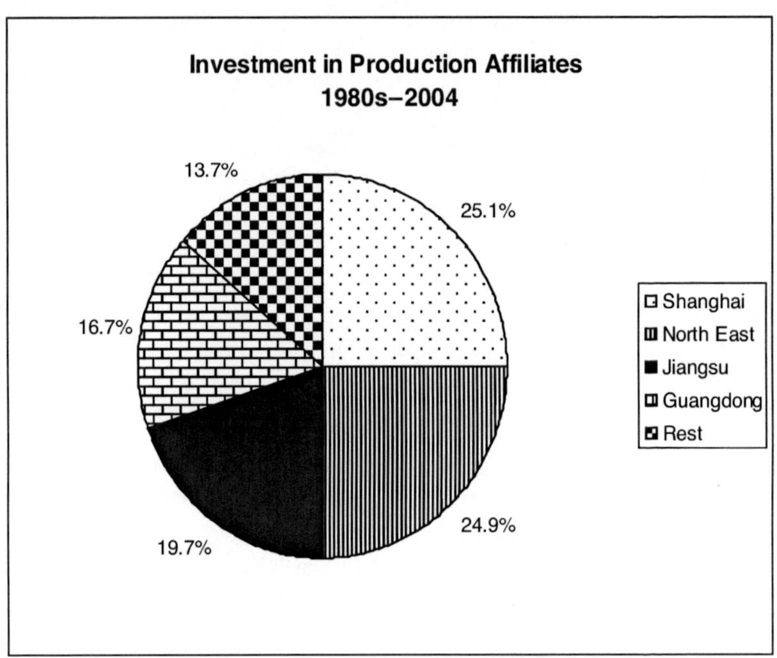

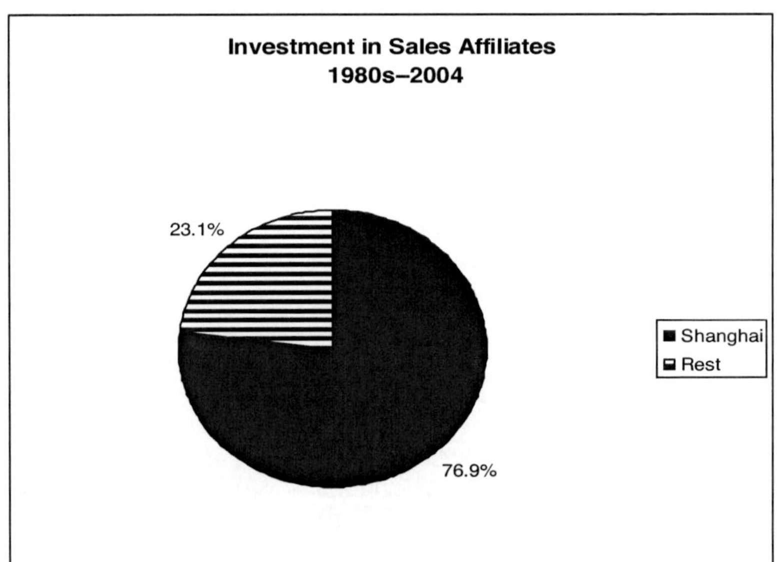

household electrical and toy industries sprouting up in Hong Kong. The process and logic were then replicated with a plant in Taiwan in 1969. Subsequently, as Hong Kong's wages rose and China started to open up by the end of the 1970s, Mabuchi was one of the first on the consignment trail in Guangdong, controlling from Hong Kong and manufacturing in that adjoining province which had become the testing ground for accommodation with the capitalist beyond. The Dalian plant opened in 1988, however, had a different set of references. It produced more complex motors for automobiles, printers and cameras, and at the time these motors were all destined for major assemblers back in Japan (Hodo and Xie, 1995). Mabuchi's international division of labor had posited China in Japan's manufacturing supply chain. But China itself was at this juncture becoming part of the act. Hence came the Jiangsu plant, set up in 1994 by Taiwan Mabuchi. Here the unequivocal target was the Chinese market, serving the makers of audiovisual devices, automobiles, vacuum cleaners, hair driers, pencil sharpeners and all the rest who are now pumping the sinews of China's growth. Finally, in 2002 Mabuchi inaugurated a sales affiliate in Shanghai. For Mabuchi, China is now a manufacturer and a market on a global stage.

And this is what we are ultimately after here. What has China come to mean for a wide range of Japanese specialist medium-sized

Figure 13.2   The Dominance of China: 2004–2007

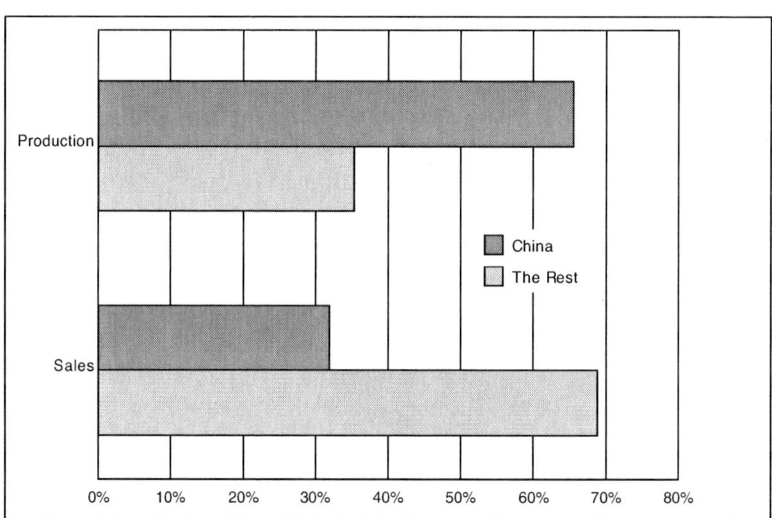

companies, and have there been any signs within the short period from 2004 and 2007 which can help us to formulate some tentative conclusions? Figure 13.2 shows the percentages for new affiliates established by the population of 429 LMEs with respect to China on the one hand and the rest of the world on the other. In the case of production affiliates, the LMEs set up 136 in China and 73 elsewhere. That works out as 65 percent and 35 percent respectively. In other words, during this short period these Japanese LMEs were setting up *over two thirds* of their overseas manufacturing units in just one country, China. As far as sales affiliates are concerned, not surprisingly China took in less than half, its figure being 45 as opposed to 98 for the rest. However, as we shall see below, designated sales affiliates are on the rise in China, and it still led the rest of the world, recording over twice as many as the next in line, the United States.

Turning to how the distribution mapped out in China itself, Figure 13.3 shows that in the case of production both Guangdong and the Northeast have remained consistent and virtually unchanged in terms of the percentage of LME affiliates accommodated. But the other three slices indicate considerable change. For one, Shanghai has shrunk drastically even in that short period from just over 25 percent to under ten percent,

Figure 13.3   LME Affiliates Established in China, 2004–2007

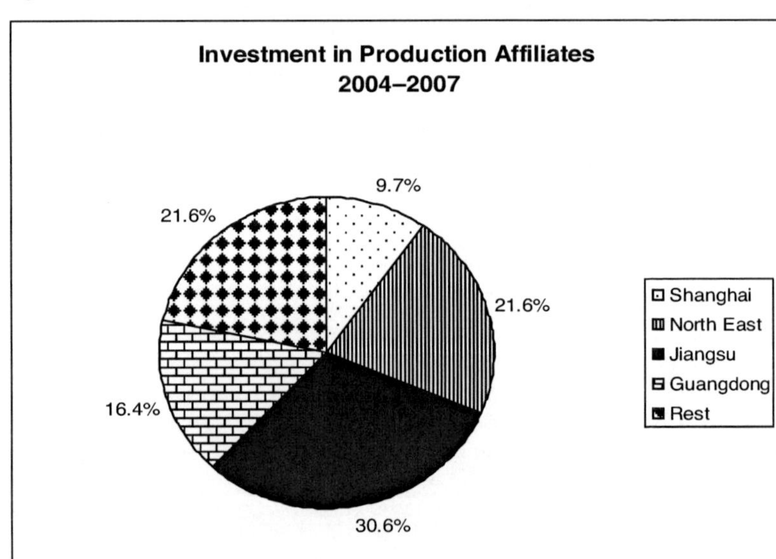

clearly suggesting that it is fast becoming too expensive and congested for factories. But it has hardly been abandoned; plant construction has simply moved out to the more spacious and less expensive adjoining province of Jiangsu, the big gainer. This movement outwards is then further emphasized by the substantial increase indicated for the rest of the country; new provinces are beginning to come into vogue, notably Hubei and the city of Wuxi within it. So production is beginning to be spread around. But this cannot be said for sales headquarters; Shanghai's percentage there is virtually unchanged. What is more, the proportion of sales affiliates to production affiliates is on the rise and the overwhelming choice is Shanghai, especially for companies who opt for sales affiliates as their initial entry route into China.

The commitment has been made. The configuration is taking shape. Regional integration is extended across the Sea of Japan to the great port and commercial center of Shanghai, from whence to manufacturing locations dotted along the coastal areas both north and south and now nudging tentatively inland. And yet during the same period of activity involving a growing number of these LMEs, Japan's business community has been assailed by reservations which have spawned alternative approaches, while at the same time new markets have emerged to present fresh opportunities. The reservations come from the fact that power outages are common and the water supply precarious; salaries are rising fast for the essential high-grade engineers and managers; there is the constant danger of knowhow leakage with scant recourse to legal compensation; Chinese local governments are continually rolling back special tax and other privileges for foreign investors relative to indigenous firms. Meanwhile, looking a few years down the road there is a mounting awareness that the merit of low production costs, especially entailing cheap labor, can disappear along the way. Moreover, increased automation and processing methods make it more viable to return manufacturing back to Japan as alluded to in the chapter on *monozukuri*.

Large corporations are taking the lead here as illustrated by camera and printer maker Canon; having shifted back the production of inexpensive cameras from China in 2004, in 2007 it established a base for the development of low-cost production in the city of Kawasaki just outside Tokyo (Nikkei, 2007j). Alternative locations are also on the agenda. According to JETRO chairman Hayashi Yasuo, last year total investments by Japanese companies into ASEAN exceeded investments into China for the first time since 2002 (Khanna, 2007). The mood among LMEs is reflected by Mizuno. Looking to yet another Olympics – this time in Beijing in 2008 – Mizuno announced in 2005 that it was aiming to double its number of outlets in China to 1,180 (Nikkei, 2005e). By 2007 it had built up a situation where 80 percent of its sports shoes are currently manufactured in China. However, personnel costs were rising there and it was facing mounting concerns about pollution, so Mizuno has now decided to spread the risk by moving some of that production to Vietnam (Nikkei, 2007k). But note that, as is now often the case with Japanese firms, the risk is being spread *within* the 'domestic' global region of East Asia.

And then there are the emerging economies of India, Eastern Europe and Russia. Increasingly, the Japanese specialist manufacturer is being presented with opportunities to replicate its East Asian regional experience on a truly global scale. China now has a definition, albeit increasingly under scrutiny, in the eyes of many a Japanese LME. The emerging economies are next on the list for investment so it is said, but little action has backed up the words so far. Do they have the capacity to be physically present in all these places? Or, at least as much to the point, given their advanced expertise coupled with a shrinking world, do they need to be present in all of them? These are the questions that Japanese LMEs will be addressing over the coming decade and for many of them China will be a very pertinent reference.

Summary

It is possible to discern portents as to what is in store for Japanese LMEs from events as they have unfolded during the brief period from 2004 to 2007. These can be viewed from a number of standpoints including stock exchange status and merger and acquisition, for example. However, there are two factors likely to affect LMEs significantly over the next decade:

1.   *Ownership patterns*: Both the numbers and stakes of foreigners are rising. Foreign institutional investors are becoming more visible and aggressive, taking higher stakes. Investment funds, as 'shareholder activists', are making the most noise, and Steel Partners the most of all. Merger and acquisition does not as yet have deep roots in Japan, but the takeover as a practice is now where the battle line is drawn. Japanese firms are reacting by putting up defenses in the form of poison pills and these measures are more often than not supported by their AGMs. The government and stock exchange may imply disapproval but the courts are still sanctioning poison pills as with the case of Bull-Dog vs. Steel. It is argued here that by concentrating on one stakeholder with one objective, funds threaten to stifle *monozukuri*.

2.   *China*: For most of the 1990s the Japanese were aware of China but not unduly alarmed by it. But this suddenly changed as its export power became apparent. Here was a vast new wave of *monozukuri* spearheaded by the 'new' China. It now has its own crop of highly successful entrepreneurs. Having opened up to catch the global surge of FDI, China has been complementary to information technology and modularization, while its cutthroat competition at home has undercut competition abroad. However, it is also a huge importer of intermediary goods and machinery for reasons that its development model is still mainly based on unskilled labor and it has a paucity of technical and managerial talent. Short on the ability to develop home-made *monozukuri* it is mainly at the bottom of the 'smiling curve', so it needs the components and equipment Japanese LMEs are in a position to supply. To 2004, investment by LMEs in production facilities was fairly evenly distributed along the coastal areas, but Shanghai took three-quarters of the sales affiliates. From 2004 to 2007, China was host to *two-thirds* of all Japanese LME investment in production affiliates. Having made a strong commitment, however, Japanese firms are looking more closely at China's shortcomings. It will be a pertinent reference for future moves by LMEs on the global scene.

14
Conclusion

The achievement

The specialist manufacturer is now a global reality, or to be more precise it is a phenomenon which, having first seen the light of day on the thoroughfares of early commercial concentration, subsequently diversified via the multitudinous byways of industrialization *en route* to venturing along the path of globalization. Generically speaking it is of universal provenance, becoming increasingly prevalent with economic advancement. The emphatic grooves of its fastidious calling have tended to constrain growth within an optimum, manageable scale of operations. Never static and ever conscious of its partnership with change, the specialist firm nevertheless associates its competitive advantage with its sustained accumulation of resources amounting to a stubborn rear-guard resistance to their unreasoned realignment. Change must have its own motivating justification. Adhering to this formula, the best of the genre have been able to carve out an enviable status for themselves in the markets they pursue, maintaining their position by constant improvements in quality and service.

Claims to universal characteristics aside however, up until very recent times at least the specialist manufacturer – even if 'born global' – has been indisputably the progeny of a particular national environment. Not simply the creature of economic determinism, it has emerged from a defining history and culture acting to shape singular national institutional structures. These structures and how they have evolved have an enduring effect on the form and behavior of the firm. This manufacturer's rules and procedures, the technologies it handles, the proprietary knowledge it acquires, whom it deals with and how it conducts its affairs are all influenced by the structural backdrop, which in modern times has almost invariably in the first instance been national in character. Japanese specialist manufacturers are no exception. Moreover, the fact that many of them can also be designated by a unique reference, *chuken kigyo* or leading medium-sized enterprise, is testimony to the special place they have in that particular national economy.

In attempting to depict the interplay of institutional forces giving rise to the LME in Japan, that nation's dynamic has been portrayed here as a montage of conflict and cooperation generated by, on the one hand, the 'samurai' representing the elite establishment and central power and, on the other hand, by the 'artisan' representing the commoner and the parochial. Having once been historical realities, even as abstract concepts these characterizations remain valid as explanatory tools for shedding light on a Japan, far less homogeneous and consensual than is all too often claimed, as it has industrialized and modernized. Looked at from this perspective, it is possible to conjure up an explanation of how things evolved from the late 19th century to the 1940s starting from a standpoint of what could be termed 'universal common sense' and then capping that with a Japanese overlay.

To take universal common sense first. Uncompromisingly opposed to the menace of western domination, the samurai took it upon himself to recharge Japan with the indispensable institutional accoutrements of the modern industrial state. This of necessity included large commercial and manufacturing corporations which, for reasons of convenience, savings on expenditure and flexibility in operations, availed themselves of smaller suppliers. Come the 1920s, the government was increasingly conscious of the need to upgrade the quality of what the latter produced and took various measures to encourage and even coerce this eventuality. Not about to look a gift horse in the mouth, these smaller companies took advantage where they could. However, the artisan also valued his freedom, just as the samurai prized his authority. As much as he could, the capable artisan, and perchance his putative LME, looked to staying outside the loop of ultimate samurai control, and he did this not least by honing his specialization.

The attractions of autonomy and the temptations of power thus alluded to are universal. But now we shift to the Japanese overlay embodied in the nation's distinctive ethical code and its tendency for hierarchical organization, which are interlinked. The ethical code is epitomized under the rubric of trust. However, this is not a trust predicated on the collaboration of more-or-less equals interacting on a tolerably level playing field, but rather a compromise based on the mutual understanding of the prevailing hierarchy. That hierarchy having been recognized, ethical propriety dictates the give and take of reciprocal obligations. A cardinal ingredient of the ethical code, especially in the traditional Japanese interpretation of Confucian teaching, is loyalty, which was presaged again on vertical association: *zaibatsu* and subcontractor, father and son. Presented thus, loyalty is anticipated both *between* institutions and *within* them. With respect to the latter, however, the personalized and all-embracing nature of loyalty strains the bounds of meaningful human contact and limits its reach.

By contrast, one of the achievements distinguishing Japan from other non-western countries is the headway it had made by the 19th century in depersonalized social arrangements not beholden to kinship ties. This enabled greater organizational size and complexity already evident in its large trading houses at the time, while it subsequently facilitated adaptation to the demands for largeness inherent in modern industrialization. The samurai duly responded. But there was always the artisan tugging at the more primal Japanese overlay; outsourcing was self-evident because of the high degree of trust, smaller units were desirable so that tangible loyalty could be assured. These were national forces which contributed to the shaping of the Japanese LME. That having been said, there was just one more universal sentiment shared mutually by both samurai and artisan, starting with the rude awakening in the mid-19th century and reasserting itself with a vengeance after the war: the all-consuming desire to 'catch up' with the West.

Herein was the field of tension which galvanized the high-growth surge from the second half of the 1950s through the 1960s. Both samurai and artisan were determined to show what Japan was made of. In order to succeed in this speed was of the essence, not only to make up lost ground but also to keep pace with accelerating world developments, as well as to cope with mounting external demands for opening up and liberating the economy. From the samurai's perspective, for speed to be sustained, choices had to be made and objectives determined. The result was initially an industrial policy top-heavy with the priorities of growth, favoring the capable, especially in the targeted industries. Inevitably in such a pressurized atmosphere, competition was fierce; many were to stumble, a select few were to scale the heights of national and eventually global recognition. The targeted industries included the likes of automobiles, machine tools and electrical appliances, where specialist manufacturers and parts suppliers had already established themselves over the previous decades and they were now to be joined by a hefty new influx. Early on the key to success was a capacity for mass production, an emphasis which would be superseded over time by demands for more intricate specialization.

Moreover, this was an era where what came first was the hardware, the traditional domain of *monozukuri*. Thus were the preoccupations of the samurai locked into the predilections of the artisan. For although many an entrepreneur was doubtless not averse to fame and fortune, the lore of *monozukuri* is the lore of making things, not, in the first instance, making money. And that entrepreneur is the scion of a national ethos which elevates the making of things for the benefit of others above the making of money for mere self-enrichment. So to revert back to our notion of universal common sense, it is not difficult for anyone to understand that a nation bruised by defeat would resort to whatever means it could to pick itself up at a fast clip. However, the degree to which the Japanese specialist manufacturer has succeeded is every bit as much derived from attitudes embedded in his society and culture. As a consequence, for specialist LMEs takeover was not an imminent menace; they remained compact and focused while being afforded space to build their own networks and do their own thing. Again, the Japanese overlay is an indispensable element of the explanatory mix.

But the extent to which this was pertinent was to be challenged and modified by increasing contact with the world beyond as the 1970s took over. Not that all was inimical by any means. The demand for miniaturized devices and energy-saving solutions stimulated by the two oil shocks, for example, fitted neatly into *monozukuri*'s remit and proved to be a happy hunting ground for LMEs hungry to make their mark with proprietary brands. Furthermore, the mounting insistence on the supremacy of knowledge was hardly something the budding LME would shy away from, any more than the incorporation of increasingly complex and ubiquitous software into the hardware frame. Actually these were stimuli for go-ahead specialists increasingly detached from government oversight and any vestiges of industrial policy. The very nature of the LME character meant that they were already pursuing a course of selecting and concentrating, a survival strategy which would become more and more *de rigueur* as the world impinged. At the same time, this approach still rested on a home base, not planned but nevertheless structured by differential institutions singular to Japan, even as catching up and the attendant policy logic were running out of steam. Extensive and overlapping cooperation between firms of all sizes still operated as did the belief in the virtues of long-term value creation. The Japanese specialist LME had grown up with and continued to anticipate relying on these props, the autonomy and success of the individual firm notwithstanding.

Where modification came in, it was to cope with the phased dilution of the national context in which these LMEs had evolved over the third quarter of the 20th century and with the collapse of the neat regional pecking order where Japan and its emerging industrial might had early on called most of the shots. As large clients shifted aspects of production offshore they had to be followed or otherwise accommodated, just as domestic costs were rising and the interest in old-style *monozukuri* among the pool of potential young recruits was declining. Added to which, as neighboring economies took off and grew, they – especially the Koreans, the Taiwanese and the Chinese – were not content to play second fiddle when they could do it themselves. Then, to cap it all, through the 1990s and into the 21st century, Japan underwent a prolonged recession which took a serious toll on its self-esteem as an industrial powerhouse. And yet most of its specialist manufacturers stood their ground and retained their distinctive identities. Moreover, most of the modifications amounted to a positive engagement with the globalizing environment, manifested particularly by increased exports and constructive FDI. This is the achievement of the Japanese specialist manufacturer and LME and the industrial structure that underpins it. Having established and learned to ensure a foothold in their national economy through the development and elaboration of carefully defined core competences, the LME took on the outside world by adapting to new practices and structures while not compromising on its *monozukuri*, illustrating in its own fashion into the bargain that globalization functions on various scales and with differing characteristics.

The question

The question is whether this can be sustained. This amounts to asking what is worth keeping and how, and what novel approaches are necessary while also being viable. What is worth keeping has been the leitmotif of this book: the medium-sized specialist manufacturer committed to excellence and the intensive development of a tightly circumscribed product range. The problem lies in the details. For one, an uncompromising pursuit of excellence is put at risk if there is a shortfall in expertise, experience and skills. And unfortunately for Japan, demographics and educational preferences are indicating this apparently inevitable eventuality. At one end of the life cycle postwar baby-boomers are retiring, at the other the birthrate has been recording historical lows for two decades. Approaching the middle students are shunning engineering and even science in general in droves, while right in the middle it is now not uncommon for at least half the skilled workers in a firm to be over 50. Solutions range from *monozukuri* colleges to the rehiring of retirees and refurbishing on-the-job training programs. Then there are two more solutions which are complementary to each other. The first is the application of information technology to record accumulated knowhow on data bases. This in turn increasingly facilitates the second solution which is the transfer of manufacturing and processing to overseas locations, particularly in the case of Japanese LMEs to China and the rest of East Asia.

However, the more this happens the more the Japanese LME will be obliged to release itself from the national institutional framework bequeathed by the path dependency process from which it originally emanated. This may not be easy despite the fact that, as has been apparent, LMEs and especially their founders have often been mavericks in their home environment, bucking the mainstream prejudices, pushing uninvited through the door. But mavericks can get tired of status impoverishment and if they play their cards right they can settle down, basking in the approval of that very institutional framework they have briefly taken on. When they then venture abroad they are likely to carry with them its basic assumptions. The international marketplace is a struggle between national institutional frameworks and their inherent assumptions of how to do business and Japan has not done badly at – again notably in East Asia – imprinting some of its own on that marketplace. There are definite limits to this though. Even in pluralistic North America where diverse business practices are easily accepted, the Japanese LME can hardly expect to enclose itself in a virtual replication of its native framework. Increasingly this LME's key to survival is a phased disentanglement from its more idiosyncratic and exclusivist tendencies. It should be preparing itself for the day when its non-family president is 35 and over half of its employees are foreigners for example.

Some discerning Japanese observers wonder, though, about their fellow countrymen's capacity for response and adaptation. With respect to the LME, we can address this aspect from the standpoint of the chamber as elaborated in Chapter 4. In fact, in the early days, with its compact modality this chamber proved an ideal vehicle for catching up and servicing an economy possessed of ordered transaction channels, and as a vehicle for nurturing and developing core competences it still is. Indeed, as time moved on, many of these specialist manufacturers went beyond catching up and in the process extended their markets beyond the narrow channels they had commenced with. In other words, they proved adept at changing gear from catch-up mode in order to carve out leadership positions in their chosen métiers. But, suggest the critics, too much success in one direction threatens to blinker the vision. Strengths which were once dominant are dwindling in importance while insufficient attention is being paid to what needs to be done. Differentiation purely by technology no longer suffices. Item-by-item *monozukuri* in which the Japanese have excelled is seen as the primitive progenitor of what should be transpiring in an information age where the marketing is as essential as the making and the foundation of production is shifting in emphasis from the hardware to the software. That is to say, if *monozukuri* is to continue to serve a purpose as a mobilizing totem it has to learn to fly. As we have seen in Chapter 10, the samurai administration has already cottoned on to this. The problem lies in the creaking hinges of institutional reform.

For the LME on the other hand, this airborne *monozukuri* calls for not abandoning the chamber as it stands but propelling it along a high-technology trajectory coupled with an extensive accommodation with external business and technology management resources in order to accelerate response and adaptation, including technology integrators and funds. Flexibility and imaginative cross-fertilization are what are needed. Pertinent to this is the role of the individual. In comparing themselves to the Chinese, the Japanese claim that, whereas the Chinese win out on a one-on-one basis, the Japanese reign supreme group for group. One could go on to surmise that perhaps this was why the Japanese caught up with the West much faster than the Chinese. The catching up having been achieved, however, there could well be less need for such pervasive group solidarity. For project-oriented work scanning the unknown the talented individual should be released from obligations to conform while respecting the ultimate need for harmony and mutual respect within the company. Arguably, because of its particular characteristics which set it apart, the Japanese LME is well primed for this. Indeed, not a few of the founders introduced in these pages have been just that because of their strong individual personalities. It will be interesting to see to what extent the national temperament is inclined to bend and how much the corporate psyche is prepared to embrace a diffusion of individualistic traits within its ranks. And the larger question restated is: can the Japanese LME modulate its own character and in so doing further extricate itself from the confining strictures of national attitudes and institutional forms which, in their entirety, are increasingly ill-suited to the demands and pressures of globalization?

The potential

Japanese leading medium-sized enterprises have room for maneuver here. This is because they are on the right side of a global trend. They are specialists. True, there are oligopolies in certain industries like tele-communications and automobiles which are consolidating further. But these oligopolies are in fact the focal points of networks. Achieving such a status, moreover, paradoxically places limitations on what oligopolies actually do for themselves. Vertically integrated manufacturing is giving way to networks where extensive use is made of independent suppliers who possess increasingly complex technologies which can render skills non-transferable and enterprise-specific. Certain fields like scientific instruments and electronics offer scope for venturesome entrants. Generic technologies make of intermediate products multi-purpose inputs extending over a range of industrial sectors and combining with other technologies from a variety of sources. The real profits are not in the washing machine but the machine tools, the components and the after-service.

This trend has filtered up to the diversified generalists, like the large Japanese electrical appliance manufacturers, as they begin to shed what is not strictly relevant to their perceived key core competencies. They too may well see their real profits in their components businesses. In other words, they are now taking their cue from the specialist LME, rather than the ambitious LME attempting to follow them as could have been the case in the 1950s. So, although this book has implied that there is an optimum size for many specialized firms, even more pertinent is focus. And focus is the word not only for the intermediate and industrial supplier but also for the many manufacturers of end-user goods, for which the markets are too small or too intricate for oligopolies to contemplate. Focus translates into depth of market knowledge, more effective market planning, and the ability to skew the sales pitch to non-price factors. It accounts for the fact that many Japanese LMEs have cracked the elite markets of North America and Western Europe in the global arena.

It has also made of them indispensable regional actors within the context of globalization. The region in this case is, of course, East Asia, which in its totality is an informal bloc of national economies. As such it mirrors the global drama in a regional setting where the practices and values of firms and institutions with their nationally-embedded origins compete for recognition, while at the same time building up networks of disaggregated competencies aimed at a regional division of labor presaged on global reach. In this process, technologies are being internationalized at the regional level. However, as in the world at large this is hardly the reciprocal transfer of knowledge and skills along a horizontal plain of comparable capabilities. Nor has it resulted in the main in the narrowing of the technological gap. For one, Japan had nearly a century's start in the accumulation of technological and managerial knowhow over South Korea, which in turn was decades ahead of the likes of Malaysia.

What this meant was that when the call came to industrialize these countries were not prepared for producing everything themselves nor were they granted the time and the leeway enjoyed to quite an extent by Japan in the 19th century to make good the missing link comprising supporting industries. On the other hand much of this missing link could be substituted for by imports, notably from Japan. The large Korean conglomerates, the *chaebol*, could make inroads into the glamorous global industries of automobiles and electronics largely indifferent to the paucity of components and machinery of domestic provenance. Japanese firms obliged, many of them specialist LMEs. By the time it was Malaysia's turn, even the electronics industry which has driven its development is not of its own making; it is heavily American, Japanese and Korean. To boot, much of the 'local content' from supporting industries emanates from subsidiaries of foreign firms, again a fair number being Japanese LMEs.

Then came China. With its massive clout and its broad-based technological foundations, there is no likelihood that China will find itself beholden to the outside world to nearly the same degree. It has added a new dimension to the regional mix. China has its own components, machinery and materials industries which are increasingly capable of satisfying the stringent standards of foreign affiliates as well as domestic companies. And yet, as we have seen in Chapters 12 and 13, Japanese specialist manufacturers are still pouring into China, if not for production at least for sales. Why is this so? Obviously it is partially due to the need on their part to be there, in this young, vast economy as it races to maturity. But the evolution of globalization itself offers a further insight. In acquiring its own practices and institutions, globalization is in the process of assuming its own path dependency, as is regional development within it. Path dependency, therefore, is no longer a phenomenon confined within national borders. In fact, a particular East Asian path dependency relating to industrialization and modernization was emerging as early as the mid-1960s. Japan was where it had started. It had nurtured a full range of industries. Having been spawned in a competitive and demanding environment, the firms involved – many of them already recognized as leading specialized manufacturers in their own country – were exporting in the 1950s and commencing to invest overseas come the 1960s. Although, much of this investment went to where the big markets in the West were located, especially the United States, a trickle became evident in East Asia, which then grew inexorably into the 21st century.

Japanese LMEs were in at the beginning of East Asian postwar economic development. As such they can claim to be East Asian LMEs also, active contributors in that development increasingly serving a 'domestic' East Asian market, appreciated alike by industrialist and consumer. They have helped the industrialist to expand business and the consumer to enjoy the comforts of modern living. Their focused specialization slots into local needs, competing to be sure but not smothering. They gained enviable reputations for their state-of-the-art quality and these reputations stuck: they were the best in the region and not unusually in the world. Some idea of the depth and consistency of this involvement as participants is suggested by their sequential incursions into Taiwan, as cited in Chapter 12. They have accelerated regional progress by presenting expertise in a timely fashion to their neighbors. The outcomes of this expertise have then traveled from Thailand, the Philippines, China and the rest to the four corners of the world. This is what the Japanese specialist manufacturer has done and this is where its potential continues to reside.

References

Note: For Japanese-language publications written by Japanese the surname comes first; for English-language publications written by them it comes last.

Arase, David (1995), *Buying Power: The Political Economy of Japan's Foreign Aid*. London: Lynne Riener Publishers, Inc.

Baldwin, Leslie T. (2006), 'Expansion Begins at Koike Aronson', *Arcade Herald*, August 10.

Bamberger, I. (1994), 'Developing Competitive Advantages in Small and Medium-Sized Enterprises', in I. Bamberger (ed.), *Product/Market Strategies of Small and Medium-Sized Enterprises*. Aldershot, UK: Avebury.

Bartels, Frank L. (2004), 'The Future of Intra-Regional Foreign Direct Investment Patterns in Southeast Asia', in Nick J. Freeman and Frank L. Bartels (eds), *The Future of Foreign Investment in Southeast Asia*. London and New York: Routledge Curzon.

Bouissou, Jean-Marie (2003), *Quand les sumos apprennent à dancer: La fin du modèle japonais* (When the sumo learn to dance: the end of the Japanese model). Paris: Editions Fayard.

Calder, K. (1988), *Crisis and Compensation: Public Policy and Political Stability in Japan*. Princeton, NJ: Princeton University Press.

Carr, E.H. (1961), *What is History?* Harmondsworth, Middlesex: Penguin Books.

Casson, Mark (1990), *Enterprise and Competitiveness: A Systems View of International Business*. Oxford: Oxford University Press.

Chandler, A. (1990), *Scale and Scope: The Dynamics of Industrial Capitalism*. Cambridge, MA: Belknap Press.

Chemical Heritage Foundation (2005), 'Chemical Heritage Foundation and Pittcon to Present Pittcon Heritage Award to Masao Horiba', on http://chemicalheritage. org/press/pr2005/pr_05_nov04.htm

Dicken, Peter (1992), *Global Shift: The Internationalization of Economic Activity*, Second Edition. London: Paul Chapman Publishing Limited.

Doane, D.L. (1998), *Cooperation, Technology, & Japanese Development: Indigenous Knowledge, the Power of Networks, and the State*. Boulder, CO and Oxford, U.K.: Westview Press.

Dore, Ronald (1965), *Tokugawa Education*. Berkeley, CA: University of California Press.

Dore, Ronald (2000), *Stock Market Capitalism: Welfare Capitalism – Japan and Germany versus the Anglo-Saxons*. Oxford: Oxford University Press.

Drucker, Peter (1973), *Management: Tasks, Responsibilities, Practices*. New York, NY: Harper & Row.

Dunning, J.H. (1980), 'Towards an Eclectic Theory of International Production: Some Empirical Tests', *Journal of International Business Studies*, 11, pp.9–31.

Economist (The) (2006), 'Lego's turnaround: Picking up the pieces', October 28, p.82.

Economist (The) (2007a), *Pocket World in Figures: 2008 Edition*. London: Profile Books Ltd.

Economist (The) (2007b), 'M&A in Japan: Land of the rising sums', July 14, p.68.

EDB (Economic Development Board) (2008), 'Makino Asia Continues to Grow from Strength to Strength', *Singapore Investment News*, February.

Evans, Ferguson (2003), *Japanese LMEs and Internationalisation: A Study of the International Behaviour of Japanese Leading Medium-Sized Enterprises in the Context of Globalisation*. PhD Thesis.

Fortune (2007), *Global 500: The World's Largest Corporations*, pp.100–60.

Fransman, Martin (1999), *Visions of Innovation: The Firm and Japan*. Oxford: Oxford University Press.

Friedman, David (1988), *The Misunderstood Miracle: Industrial Development and Political Change in Japan*. Ithaca, NY: Cornell University Press.

Fruin, W. Mark (1992), *The Japanese Enterprise System: Competitive Strategies and Cooperative Structures*. Oxford: Oxford University Press.

Fuji Research Institute (1998), '*Monozukuri' Kakumei* (The 'Monozukuri' Revolution), Fuji Sogo Kenkyusho Sangyo Chosa Bu (Industrial Survey Department, Fuji Research Institute) (ed.). Tokyo: Toyo Keizai Shinposha.

Fujitec Co., Ltd. (1998), *Global Fujitec: 50th Anniversary*.

Gélinier, O. (1996), *La Réussite des Entreprises Familiales: Les Moyennes Entreprises Patrimoniales – Un Atout pour L'Avenir* (The Success of Family Firms: Medium-Sized Enterprises as a Trump Card for the Future). Paris: Maxima.

Gerlach, Michael L. (1992), *Alliance Capitalism: The Social Organization of Japanese Business*. Berkeley and Los Angeles, CA: University of California Press.

Gershenkron, Alexander (1966), *Economic Backwardness in Historical Perspective*. Cambridge, MA: The Belknap Press of Harvard University Press.

Hibi Tsuneaki (2002), *Shitauke Yamete: Nitchi o Mezase!!* (Stop Subcontracting: Go for the Niche). Tokyo: Wedge K.K.

Hodo Chikato and Xie Dan Ming (1995), *Chugoku de Kigyo o Sodateru Kisetsu* (Creating a Successful Company in China). Tokyo: Toyo Keizai Shinposha.

Hollerman, Leon (1996), 'Japan's Direct Investment in Britain: A Case Study of Industrial Policy on the International Plane', in Leon Hollerman and Ramon H. Myers (eds), *The Effect of Japanese Investment on the World Economy: A Six-Country Study, 1970–1991*. Stanford, CA: Hoover Institution Press, Stanford University.

Horiba Masao (2002), 'Keynote Speech', on http://www.aabi.info/event/keynote01.htm

Horiba Masao (2003), *Iya Nara, Yamero* (If You Don't Like It, Give It Up). Tokyo: Nihon Keizai Shinbunsha.

Horiba Masao (2005), *Shigoto ga Dekiru Hito Dekinai Hito* (People Who Can Work and People Who Can't). Tokyo: Mikasa Shobo.

IBO (2005a), 'Great People of Osaka: Rihachi Mizuno – Helped Sports Take Root and Flourish as a Part of Daily Life', *Osaka Business Update*, Vol. 4.

IBO (2005b), 'Great People of Osaka: Shozaburo Shimano, the World's Leading Bicycle Components Company', *Osaka Business Update*, Vol. 3.

Ikegami, Eiko (1995), *The Taming of the Samurai: Honorific Individualism and the Making of Modern Japan*. Cambridge, MA: Harvard University Press.

Ishikawa, Akira and Tai Nejo (1999), *Top Global Companies in Japan*. Singapore: World Scientific Publishing Co. Pte. Ltd.

Ito Masaaki (1996), 'Sangyo Kozo no Kodoka to Chusho Kigyo (Industrial Structure Upgrading and SMEs)', in Momose Shigeo and Ito Masaaki (eds), *Shin Chusho Kigyo Ron* (New SME Theory). Tokyo: Hakuto Shobo.

JCH (Japan Company Handbook) (various years). Tokyo: Toyo Keizai Inc.

JMAM (2001), *Kyoso Yui o Mezasu: Monozukuri Keiei Kakushin* (Aiming for Competitive Advantage: *Monozukuri* Management Reform). Tokyo: Japan Management Center, Inc.

JMIF (Japan Motor Industrial Federation) (2002), 'JMIF PR Video "The Birth of the Otomo – the First Automobile Produced Entirely in Japan" and 35th Tokyo Motor Show – Passenger Cars and Motorcycles (2001) Official Video and DVD Completed', on http://www.jama.or.jp/release/jmif/release/eng/20020304.html

Jones, A. (1987), *The Art of War in the Western World*. Oxford: Oxford University Press.

Kage Toshitaka (1991), *Yuryo Chuken Kigyo e no Michi: Sono Henshin no Susumekata* (The Road to the Superior Leading Middle-Sized Enterprise: How to Proceed with the Transformation). Tokyo: Daisan Shuppan, KK.

Keieisha Kaiho (1990), 'Kyuseicho Kigyo Shacho ga Tsuzuru: Seicho no Yorokobi to Kurushimi' (What the Presidents of Fast-Growth Companies Say: The Joys and Vexations of Growth), November Edition.

Kelly, Philip F. and Kris Olds (1999), in Kris Olds, Peter Dicken, Philip F. Kelly, Lily Kong and Henry Wai-chung Yeung (eds), *Globalisation and the Asia-Pacific: Contested Territories*. London: Routledge.

Keynes, J.M. ([1938] 1973), *The Collected Writings of John Maynard Keynes*. Vol. XIV. *The General Theory and After*, D. Moggridge (ed.). London: Macmillan.

Khanna, Vikram (2007), 'Japanese firms Back in Asean as China Loses its Appeal', *Business Times*, November 8.

Khurana, Rakesh (2002), *Searching for a Corporate Savior: The Irrational Quest for Charismatic CEOs*. Princeton, NJ: Princeton University Press.

Krugman, Paul (1995), *Development, Geography, and Economic Theory*. Cambridge, MA: The MITI Press.

KSK (*Kaigai Shinshutsu Kigyo Soran*) (various years). Tokyo: Toyo Keizai Inc.

Kuhn, Robert Lawrence (1985), *To Flourish Among Giants: Creative Management for Mid-Sized Firms*. New York, NY: John Wiley & Sons, Inc.

Kurose Naohiro (1997), *Chusho Kigyo Seisaku no Sokatsu to Teigen* (SME Policy: Summary and Proposals). Tokyo: Doyukan.

Liker, Jeffrey K. (2004), *The Toyota Way: 14 Management Principles from the World's Greatest Manufacturer*. New York, NY: McGraw-Hill.

Mabuchi Motor (1999), *Mabuchi Motor Company Profile*. Matsudo-shi, Chiba-ken: Mabuchi Motor Co., Ltd.

Marshall, Alfred (1920), *Principles of Economics*, Eighth Edition. Basingstoke and London: Macmillan Press Ltd.

Matsuura Motoo and Okano Masayuki (2004), *Gijutsu de Ikiru! Hitori Ichi Oku En Uriageru Keiei* (Living by Technology! A One-Man Business Making One Hundred Million Yen). Tokyo: Bijinesu-sha.

Ministries of Economy, Trade and Industry, Health, Labor and Welfare, and Education and Science (2004), *Monozukuri Hakusho* (White Paper on Manufacturing Infrastructure). Tokyo: Gyosei K.K.

Mintzberg, Henry and James A. Walters (1985), 'Of Strategies, Deliberate and Emergent', *Strategic Management Journal*, Vol. 6, pp.257–72.

Missions Economiques (2006), *Revue Asie: Comment l'Asie Tire Profit de l'Émergence Chinoise* (Asia Review: How Asia is Profiting from the Emergence of China), Mission Economique de Pékin, No. 132, May.

MKB (*Mijojo Kigyo Ban*) (various years). Tokyo: Toyo Keizai Inc.

Morishita Tadashi (1996), 'Kigyoka Seishin to Benchya Bijinesu' (The Enterprise Spirit and Venture Business), in Momose Shigeo and Ito Masaaki (eds), *Shin Chusho Kigyo Ron* (New SME Theory). Tokyo: Hakuto Shobo.

Morris-Suzuki, Tessa (1994), *The Technological Transformation of Japan: From the Seventeenth to the Twenty-first Century*. Cambridge: Cambridge University Press.

Nakamura, Hideichiro (1990), *Shin Chuken Kigyo Ron* (New Leading Medium-Sized Enterprise Theory). Tokyo: Toyo Keizai Shinposha.

Nikkan Kogyo (2003), *Kono Bunya Ichiban Kigyo* (Top Enterprises in this Field). Tokyo: Nikkan Kogyo Shinbunsha, pp.23–5.

Nikkei (2005a), 'Takara, Tomy Togo e' (Takara and Tomy to Unite), May 12.

Nikkei (2005b), 'Watashi no Nigawarai: Horiba Masao, Horiba Seisakusho Saiko Komon' (Cause for a Bitter Smile: Horiba Masao, Supreme Counsel, Horiba, Ltd.), Nihon Keizai Shinbun, September 19.

Nikkei (2005c), 'Obara: Zenki mo Keijo Rieki Saiko' (Obara: Again Highest Profits for Previous Term), November 22.

Nikkei (2005d), 'Chiho no Jitsuryokuha' (The Regional Power Houses), July 6.

Nikkei (2005e), 'Supootsu Yohin, Kaigai e Jikuashi' (Sports Goods, Emphasis on Overseas), June 14.

Nikkei (2006a), 'TakaraTomy to Eiga Seisaku: Indekkusu, Tatsu No Ko Puro ni Shusshi' (To Make Movies with TakaraTomy: Index Invests in Tatsu No Ko Productions), June 27.

Nikkei (2006b), 'Kokoro no Tamatebako: Horiba Masao, Horiba Seisakusho Saiko Komon' (Treasures of the Heart: Horiba Masao, Supreme Counsel, Horiba, Ltd.), Nihon Keizai Shinbun, May 8–13.

Nikkei (2006c), 'Kosaku Kikai: Kakkyo no Yukue, Akiguchi ga Shoten' (Machine Tools: Direction of Activity, Focus on Early Fall), August 8.

Nikkei (2006d), 'MBO Senmon Fando: Beikei Ote Vestar, Nihon Shinshutsu' (MBO Specialist Fund: U.S. Big Name Vestar Setting Up in Japan), August 22.

Nikkei (2007a), 'Riken, Bunsan Seisan: Saigai Risuku Teigen' (Riken Dispersing Production: Reducing Disaster Risk), July 22.

Nikkei (2007b), 'TakaraTomy – Bei Fando Shusshi Ukeire: Hokubei Jigyo Nohau Katsuyo (TakaraTomy – Accepts U.S. Fund Investment: Applying North American Business Knowhow), March 7.

Nikkei (2007c), 'Shiberia Tetsudo de Yuso: Nissu o 4 wari Tanshuku' (Transport by Siberian Railway: 40 percent Saving in Number of Days', July 20.

Nikkei (2007d), 'Toyota, Junrieki 32% Zo; Uriagedaka, GM Nuku' (Toyota's Net Profit Up 32%; Overtakes GM in Sales), August 4.

Nikkei (2007e), 'Shinko Koku ga Suijun Oshiage: Kosaku Kikai no Juyo Kocho' (The New Economies Force the Pace: Machine Tool Demand Upbeat), February 26.

Nikkei (2007f), '2006 Nendo no Mochikabu Hiritsu: Gaikokujin Saiko no 28%' (Shareholding Ratios for Fiscal 2006: Highest Ever for Foreigners at 28%), June 16.

Nikkei (2007g), 'Gyoseki "Yoku naru" 53%; Tekitaiteki Baishu "Hiteiteki" ga 7 wari' (For 53% Business 'Going well'; 70% 'Negative' about Hostile Buyouts), August 6.

Nikkei (2007h), 'Burudokku 9 gatsu Chukan: 19 Oku Yen no Saishu Akaji' (Bull-Dog September Mid-term: Final Deficit of 1.9 Billion Yen), November 17.

Nikkei (2007i), 'Kishimi o Koete: Mamoritakatta Hataraku Hokori' (Surmounting the Friction: Desire to Protect a Pride in Work), October 11.

Nikkei (2007j), 'Teikosto Seisan Rain Kaihatsu e: Kiyanon ga Shin Kyoten' (Developing Low-cost Production: New Base for Canon), March 11.

Nikkei (2007k), 'Betonamu de Honkaku Seisan: Chugoku Shuchu Minaoshi' (Production in Vietnam Gets Going: Concentration on China Being Revised), October 18.

Nikkei Business (2002), *Chugoku ga 'Sekai no Kojo'* (China is the 'World's Factory'). Tokyo: Nikkei BP Sha.

Nikkei Kin'yu Shinbun (2002), 'Yunion Tsuuru: Konshu no mo Kaigai IR: Oshu no Kikan Toshika Taisho' (Union Tool: Overseas IR this Autumn Again: Targeting European Institutional Investors), April 16.

Nikkei Sangyo Shinbun (1982a), 'Kyokusho Choko Doriru Sekai Kyokyu: Shitauke kara Dappi' (Supplying the World with Extremely Small Super-Hard Drills: Shedding the Subcontracting Role), February 17.

Nikkei Sangyo Shinbun (1982b), 'Kyokusho Choko Doriru Sekai Kyokyu: Seihin Takakuka ni Hashiru' (Supplying the World with Extremely Small Super-Hard Drills: Launching into Product Diversification), February 18.

Nonjon, Alain (1999), *La Mondialisation* (Globalization). Paris: Editions SEDES.

Ohmae, Kenichi (1995), *The End of the Nation: The Rise of Regional Economies.* New York, NY: Free Press Paperbacks.

Okazaki, Tetsuji and Masahiro Okuno-Fujiwara (1999), 'Japan's Present-Day Economic System and its Historical Origins', in Tetsuji Okazaki and Masahiro Okuno-Fujiwara (eds), *The Japanese Economic System and its Historical Origins.* Oxford: Oxford University Press, pp.1–37.

Ozawa, T. (1979), *Multinationalism, Japanese Style.* Princeton, NJ: Princeton University Press.

Pentel (1996), *Penteru 50-nen no Ayumi* (Steps in Pentel's 50-year History). Tokyo: Pentel Co., Ltd.

Porter, Michael E. (1985), *Competitive Advantage: Creating and Sustaining Superior Performance.* New York, NY: Free Press.

Reich, Robert B. (1992), *The Work of Nations: Preparing Ourselves for 21st Century Capitalism.* New York, NY: Vintage Books.

Rocca, Jean-Louis (2007), 'The Reality behind the World's Workshop – The Flaws in the Chinese Economic Miracle', *Le Monde diplomatique*, May 11.

Sato Keita (2005), 'Seiko Taiken no Oboreta Kakudai Rosen' (The Road to Expansion Drowned in the Experience of Success), Nikkei Business, 7th Feb.

Seki, M. (2004), *Beyond the Full-Set Industrial Structure: Japanese Industry in the New Age of East Asia.* Tokyo: LTCB Library Foundation.

Shimano, Yoshizo (2005), 'Watashi no Rirekisho' (My Life Story), *Nihon Keizai Shinbun*, July.

Shimano, Yoshizo (2008), *This is My Road: Life is What You Make It.* Singapore: John Wiley & Sons (Asia) Pte Ltd.

Simon, Hermann (1996), *Hidden Champions: Lessons from 500 of the World's Best Unknown Companies.* Boston, MA: Harvard Business School Press.

Storey, David (1994), *Understanding the Small Business Sector.* London: Routledge.

Taguchi, Naoki (2000), 'Higashi-Osaka Chiiki no Torihiki Bungyo Kozo – Tayo na Juyosaki o Motsu Byora Kanagata Sangyo o Jirei Toshite' – (The Transaction and Division of Labour Structure in the Higashi-Osaka District – Exemplified by Fastener and Mould Industries with Their Multifarious Demand Sources –), in Ueda Hirofumi (ed.), *Sangyo Shuseki to Chusho Kigyo: Higashi-Osaka Chiiki no Kozo to Kadai* (Industrial Integration and SMEs: Structure and Theme of the Higashi-Osaka District). Tokyo: Sofusa.

Tanaka Kikinzoku Kogyo K.K. (1985), *Precious Metals Science and Technology: An Introduction*. Tokyo: Tanaka Kikinzoku Kogyo K.K.

Teraoka Hiroshi (1997), *Nihon Chusho Kigyo Seisaku* (Japan's SME Policy). Tokyo: Yuhikaku.

Tsuchiya Yasuo (1999), *Nihon Kigyo wa Ajia de Seiko Dekiru: Guroburu Keiei o Jitsugen Shishin* (Japanese Companies Can Succeed in Asia: A Guide to Realizing Global Business). Tokyo: Toyo Keizai Shinposha.

Tsuru, Shigeto (1993), *Japan's Capitalism*. Cambridge: Cambridge University Press.

Tsutsui, William M. (1998), *Manufacturing Ideology: Scientific Management in Twentieth-Century Japan*. Princeton, NJ: Princeton University Press.

UNCTAD (2005 and 2005), *World Investment Report*. New York, NY: United Nations Publications.

Union Tool (2007a), *Summary Report on Interim Closing for the Fiscal Year ended November 30, 2007*, on http://www.uniontool.co.jp/english/pdf/q2_070717_e.pdf

Union Tool (2007b), 'Company Profile', on http://www.uniontool.co.jp/english/profile_02.html

Wang, Peter S. (2005), 'Auto Ambitions: China's Auto-Parts King Lu Guanqiu Prepares His Company Wanxiang to Take on the World', *Asia Inc*, March.

Whitley, Richard (1992), *Business Systems in East Asia: Firms, Markets and Societies*. London: Sage Publications Ltd.

Womack, James P., Daniel T. Jones and Daniel Roos (1990), *The Machine that Changed the World*. New York, NY: Macmillan Publishing Company.

Yoshihara, Kunio (1994), *Japanese Economic Development*, Third Edition. Kuala Lumpur: Oxford University Press.

Young, A.K. (1979), *The Sogo Shosha: Japan's Multinational Trading Companies*. Boulder, CO: Westview Press.

Zhu, Linan (1993), *Zhanhou Riben de Duiwai Kaifang* (Japan's Post-War Opening to the Outside World). Beijing: Dangdai Zhonguo Chubanshe.

Index